Venezuela's Oil

VENEZUELA
OIL FIELDS
OIL PIPELINES
GAS PIPELINES
MAIN CITIES
LAKES
REFINERIES
SCALE
0 50 100 150 200 250 300 Km.
0 50 100 150 200 Miles
VENEZUELA IN SOUTH AMERICA
N
Caribbean Sea
Caribbean Sea
Atlantic Ocean
NETHERLANDS ANTILLES
Is. los Roques
I. Orchila
I. la Blanquilla
I. la Tortuga
I. de Margarita
GRENADA
TOBAGO
TRINIDAD
COLOMBIA
GUYANA
BRAZIL
BRAZIL
Amuay
Cardon
Golfo de Venezuela
MARACAIBO
CORO
Bajo Grande
CABIMAS
Sierra de Perija
Lago de Maracaibo
San Lorenzo
BARQUISIMETO
VALERA
Cord. de Merida
MERIDA
Pico Bolivar 5008m.
Barinas
SAN CRISTÓBAL
Uribante
Moron
El Palito
PTO. CABELLO
VALENCIA
MARACAY
CARACAS
Coledes
Guanara
Portuguesa
Embalse de Guárico
Apure
Arauca
ORINOCO
Orinoco
Orinoco
CUMANA
BARCELONA
El Chaure
Pta. la Cruz
CARÚPANO
Caripito
San Roque
Tucupita
Delta
CIUDAD GUYANA
CIUDAD BOLIVAR
Caura
Caroni
Paragua
Carrao
Sierra Pacaraima

Venezuela's Oil

by
Rómulo Betancourt
Translated by Donald Peck

London
GEORGE ALLEN & UNWIN
Boston Sydney

First published in 1978

ISBN 0 04 338082 4 cloth
0 04 338083 2 paper

Typeset in Great Britain
in 11 on 12 point Times
by Trade Linotype Limited

Printed in Great Britain by
Biddles Ltd, Guildford, Surrey

Prologue

by Victor L. Urquidi

Venezuela now owns her own oil, which has not been the case for over sixty years. On 29 August 1975 President Carlos Andrés Pérez placed his signature on the Organic Law which reserves for the state the production and marketing of hydrocarbons (the legal instrument by which they were nationalized), and, on 1 January 1976, the Venezuelan government formally took possession of the country's oil wealth.

One of the Venezuelan patriots who has figured very largely in the story of the nationalization of oil is Rómulo Betancourt. He has twice been head of state and has fought unceasingly to preserve his country's autonomy. He began the struggle in the thirties, and did not flinch or check his efforts when in exile because of repressive dictatorships. During periods of democratic government, and when he himself was in power, this campaign started by Betancourt became a key element in the nationalist platform of his party Acción Democrática; since 1958, when the Pérez Jiménez régime was overthrown, it has become the national ambition upheld by all political groups.

This book consists of a series of articles by Betancourt about Venezuelan oil, preceded by his speech, as honorary life senator, in the Venezuelan

Senate on 6 August 1975. In it the reader will find not only the history of a great political and economic campaign, but also the passionate arguments of a defender of the Venezuelan people and its just claims, and reasoned arguments of an able administrator of the country's oil. Since the first companies were set up at the beginning of this century, two conflicting tendencies have been at work in connexion with the oil industry: nationalism and submissiveness to external influences. Betancourt tells the story of the first attempts to put limitations on the concessions granted to foreign companies, which were soon brought to nothing by the new laws favourable to foreign interests. When he came back from exile in March 1936, Betancourt helped to focus his country's attention on the oil problem. Government fluctuated back and forth, but this black gold remained entirely in the hands of foreign capital which enjoyed enormous privileges, until the Junta Revolucionaria in 1945 established the first important taxes on the industry by decree. By 1948 this government had achieved the first collective labour contract in the industry and a new distribution of oil profits in equal shares between the government and the oil companies. After another step backwards in a period during which many concessions were put up for sale, there has been progress, under the democratic governments since 1958, towards the Venezuelanization of oil. By 1975 a balance had been struck between a strong nationalism and a new sense of international cooperation in defence of oil resources. It should not be forgotten that OPEC dates back to 1960 and was founded at the instigation of Venezuela.

The recent decision to nationalize oil is consistent with the growing awareness of the Venezuelan people, despite increasing external hostility, of what the fact that their basic wealth is in hydrocarbons implies and how it can be used as the mainstay of their economic and social development. As recently as 1974 the Congress of the USA overruled White House objections and excluded the OPEC countries, including Venezuela and Ecuador, from the tariff concessions granted to 120 nations. But the nation's will has won through in Venezuela, and has been put into practice in an effective way, so that Venezuela now owns her own wealth and can run it with the interests of future generations at heart.

Let me draw the reader's particular attention to the arguments used by Rómulo Betancourt in connexion with the controversial Article 5 of the Organic Law, which allows the Venezuelan government, after a special vote in Congress, to sign joint contracts in partnership with private firms, for limited periods of time, and under the overall control of the state.

Given the size and complexity of the oil industry, Venezuela could not afford to be hamstrung and run the risk of a technical failure. The nation's basic interests are at stake, and they hit on a solution which will ensure efficient administration without infringing the country's sovereign rights.

The essays which make up this book and the documents which are included are a very useful and necessary addition to Betancourt's great and earlier book, *Venezuela, Política y Petróleo*, first published in Mexico (Fondo de Cultura Económica, 1956).

These books give a full account of Venezuelan oil

policy in recent years, culminating in its nationalization. It is up to that new generation to go on ploughing back the profits of oil, as proposed so brilliantly by Betancourt and his followers thirty years ago, into society as a whole; in the next decade the results of this policy will become even more striking than they are now. Let this policy, which was started in Mexico in 1938, set an example for the rest of Latin America.

Contents

The Venezuelanization of Oil

In 1964, when I took my oath of office as honorary senator for life, the high position which the Constitution assigns to democratically elected presidents after their term of office, I said that I would not play an active rôle in Congress. I believe that, as a former head of state, I should keep on the sidelines of the day-to-day business of party conflict, acting as a moderating influence in my country, and offering all the help I can for the maintenance of the democratic system in which the overwhelming majority of the Venezuelan people passionately believes. Today I have come to speak here, two days after ex-President Caldera, because on such an extraordinarily important occasion as the discussion of the law which reserves to the state the production and marketing of oil, our basic source of wealth, it would be wrong for the opinions of former presidents to be lacking.

An anecdote and two observations

Let me begin with a quotation, an anecdote and two preliminary observations. The quotation is from

Davenport and Cooke, in the book they published in 1923, which caused a sensation in financial circles across the world, called *The Oil Trusts and Anglo-American Relations.* They wrote: 'It's a common saying that oil gives a free rein to the most evil passions, producing in businessmen a greed more all-consuming than gold fever and moving statesmen to Machiavellian designs.'

The anecdote was told me in London in 1965 by a highly-placed executive of an oil company with a sense of humour.

Once upon a time an executive jet carrying ten directors of big oil companies blew up in mid-air. They were Catholics and had contributed generously to the upkeep of charities and religious foundations. St Peter didn't know how much their generosity was due to the fact that the sums they donated to orphanages and hospitals reduced their income tax payments. So they got into heaven and went on talking about oil to the exclusion of everything else, until one of them, rather bored by the one-track minds of the others, brought up the idea that there might be oil to be found in hell. One after another the oil barons leaped from heaven with its saints and cherubim down to the darkness and boiling cauldrons of Lucifer's kingdom. St Peter then asked one of the remaining oilmen, who happened to be the man who made up the story: 'Why are they going to hell?' He replied: 'I was fed up with their conversation on the single subject of oil, so I suggested that there might be oil down there.' In the end they all went except this story teller. When St Peter saw that he too was about to jump, he said: 'But you know that it's a lie.' But he leaped into the darkness, saying: 'Perhaps it's true.'

My two preliminary observations serve to point out that you are going to hear not the ideas of an oil expert trained in Oklahoma or one of the other universities that specialize in the subject, but of a man who, from 1928, when he went into exile for the first time, until today, has studied with great concern and dedication the oil problem, in the belief that the economic, political, social and cultural life of our country has come to revolve around oil. In addition I have had practical experience of managing the relations between the industry and the state for two periods, first of all from 1945 to 1948 as President of a *de facto* coalition government, and then from 1959 to 1964 as a democratically elected head of state.

My second observation is that, when speaking about the multinational companies which have extracted from our country millions of barrels of oil and billions of dollars and pounds sterling of surplus profits, I will adopt a tone of voice which contains neither anger nor satisfaction. That is because I believe that those companies have only been able to take such great spoils from our country because they have found ready helpers among unscrupulous or corrupt rulers who could not or would not defend the interests of Venezuela.

Last night I gave a lecture to a large group of doctors, lawyers and economists who wanted to hear me talk about oil. I was able to show that the accumulated evidence on the history of the industry in this country is very extensive. But today I will try to be more concise.

Early exploitation of petroleum

It is well known that the exploitation of petroleum in this country had a quiet beginning when Dr González Bona, a small-town doctor from Junín in the state of Táchira, discovered that an oily sludge on the surface of the river, in a place appropriately known as La Alquitrania, was tar. González Bona was not just a small-town doctor, but also an engineer and something of a chemist, and he formed a limited liability company, to be called Petrolia del Táchira, with General Baldó and one of the González Rincón family. One of the partners in the company was sent to Pennsylvania, then the Mecca of the oil business, because it was there that the subsoil had been drilled for the first time, by an adventurer called Drake. This first company came to an end in 1912 for lack of working capital to continue operating: it had only managed to produce 60 barrels a day.

The real exploitation of petroleum began with its first cousin, asphalt. It has a very interesting story, which I won't, however, tell you today in full detail.

Asphalt was produced in the Guanoco Well in the state of Anzoátegui by the New York and Bermúdez Company. The company's concession was declared null and void because it did not comply with the conditions in its contract with the government. But the New York and Bermúdez Company went on producing oil from the well illegally. In 1899 Cipriano Castro and 'The Group of 60' came into power. The Castro régime came into conflict with the New York and Bermúdez Company, which preferred, instead of paying its taxes, to help to finance the Revolución Libertadora against Castro. A representative of this company went with General Manuel María Matos to Europe to buy the Ban Righ, and also paid contributions towards other weapons and ammunition for the revolution. The Venezuelan people snatched up these arms to fight against the Castro régime, which had come to be hated for its bad-tempered despotism. But the uprising was defeated at La Victoria. Next came the blockade by the Kaiser's Germany and by Britain, then at the height of her imperial power. Both countries then used gunboats to press for payment of interest and instalments on loans to debtor countries in default, like Venezuela. On that occasion Castro, for whom some have tried to create a nationalistic halo, was only too ready to accept aid from the United States, in the shape of two ships, the *Cincinnati* and the *Topeka,* to protect him from the blockade.

Soon afterwards began the lengthy rule of Juan Vicente Gómez. President Castro had sailed for Europe on the *Guadalupe* to be attended by his surgeon, Israel, in Berlin. But he forgot about the tendency of newly installed rulers to bite off the hand that fed them, and what the Argentinian dictator, Juan Manuel de Rosas, after being exiled for a time in Montevideo by the man he left temporarily in charge in Buenos Aires, called the *ley de la patada historica* ('dog bites dog').

The Venezuelan people will not forget the Gómez period, not only because of his cruelty towards his opponents who were faithful to their love of liberty and democracy which is a vital part of our national ethos, but also because of the way in which he hawked the country's wares to the foreign companies which exploited us. The social, economic and cultural life of the country did not just stagnate; it took several steps backwards. So Mariano Picón Salas was right in saying that 'Venezuela only entered the twentieth century in 1936'; up till then it had remained in the nineteenth century.

Gumersindo Torres

However, in the midst of a sea of corruption, in which concessions were granted by the Ministry of Development to the Gómez family, and to the favourites and friends of the government, a man appeared who, although not even an engineer but a doctor without any university training in these matters, was able to check the squandering of national resources in concessions to foreigners with the energy of a nationalist. This was the trusty Gumersindo Torres, who in 1917 took over the Ministry of Development, which then controlled all oil affairs. He at once took steps in favour of the real interests of Venezuela. He changed the lax Mining Law which had been in force since 1909; his new law fixed a maximum duration of 30 years for concessions. In addition it provided for any concession which was not exploited in the first three years after it was granted to the government. The law said that, given that oil is a typical non-renewable natural resource, half of each production area should be set aside as a national reserve, the property of the state. What is more, Torres even pressed for a petition to the Supreme Court to get it to annul the concessions of the Caribbean Petroleum Company which stemmed from the famous concessions to Dr Rafael Max Valladares and those of the Venezuelan Oil Concession Company. In order to get round Torres's intransigent defence of our country's interests, the Compañía Venezolana de Petróleo was set up at this time. This was the company of Gómez, the dictator from Maracay. In 1946 the manager of this company admitted to the Tribunal of Civil and Administrative Responsibility what could anyway be proved from its books, that Gómez had got on his own account more than 20 million bolívares for concessions granted by the company. As a result of pressure from the oil companies and from the British and US embassies, particularly the latter then in the hands of the very aggressive Mr McGoodwin, Gumersindo Torres was removed from office.

Oil legislation

In 1921 came one new law, and in 1922 another. Both tended to hawk the country's wares and were totally submissive to the oil companies and for one simple reason, they were drawn up by an oil lawyer, Dr Rafael Hidalgo Hernández. Gómez had summoned the managers of the companies and told them: 'You know all about oil and we don't; prepare a law to regulate your industry.' In other words they were to be able to have their cake, to divide it up, and then eat it. It was put this way not by an aggressive extreme nationalist but by a US journalist, Clarence Horn, writing in the quintessentially capitalistic American magazine *Fortune* in March 1949. The new law gave longer terms for all concessions, and included a new privilege which was to take on great importance during the subsequent nationalist campaign for the gradual recovery of our control over our oil. Without meaning to be paradoxical we could call the goal of this campaign one of revolutionary evolution. The all-important article in the new Gómez law gave the oil companies the right to import anything they liked through customs free of duty, which amounted to making contraband legal.

Competition for concessions

Competition between the oil companies to get concessions in our country grew more intense after the end of the First World War. As a result of the war liquid fuels acquired a supreme importance which they had not previously enjoyed. Ships and submarines began to burn not coal but fuel oil and other petroleum by-products. Aeroplanes and motorized infantry were also in urgent need of fuel produced from hydrocarbons. In this situation President Clemenceau of France, for instance, sent his pathetic telegram to President Wilson of the USA asking for 100,000 barrels of oil to be shipped to France so that he could make a stand against the Germans in the decisive battle of the Marne.

General Ludendorff, one of the leading generals of the Imperial Powers in that war, attributed the defeat of Germany and Austria to their lack of oil. When the war ended Britain's First Lord of the Admiralty, Lord Curzon, said in almost poetic terms: 'Our victory came on waves of petrol.' The events of the war radically changed the US attitude to oil. Up till then the man who built up the biggest integrated monopoly in the history of capitalism, John D. Rockefeller and his Standard Oil, concentrated exclusively on controlling oil production, refining and distribution inside his country's domestic market.

P. H. Frankle, a well-informed oil expert, in his book *L'Économie Petrolière* says that before the post-war period the only contact between Standard Oil and the government was through tax-collectors and officials of the judiciary who tried to enforce the Sherman anti-trust law. But after the war President Wilson, and his successors too, personally encouraged the oil companies to look for oil all over the world. In this way began the close interdependence between diplomacy and oil.

Frankle points out ironically that attempts by the oil companies and by the State Department to deny these links either suggests that both lacked intelligence and imagination, or reveal that they were as hypocritical as the British, who even used to say that they had built up an Empire without meaning to.

In Venezuela the British arrived first. Royal Dutch Shell, in an operation which Sir Henry Deterding once described as the greatest financial adventure of his life, bought up 51 per cent of the shares of the Caribbean Petroleum Company. But, in Venezuela as in other parts of the Western world, the dynamic dollar eventually defeated the ailing pound sterling. Standard Oil arrived last but, before very long, became the biggest producer. It obtained concessions not only in western Venezuela, but also in the east, which it pioneered in opening up.

In 1930 Gómez, perhaps as a result of one of his strange whims, brought back to his Development Ministry Gumersindo Torres, who proceeded to go on defending the real interests of our country with admirable obstinacy. He laid the foundations for the technical body to supervise the industry to which ex-President Caldera referred in his speech. In 1930 he sent a historic note to the companies, in which he attacked the uncontrolled way in which the companies imported all kinds of goods through the country's ports, and produced some very revealing figures. In the seven years leading up to 1930 exemptions from customs duties had reached a total value of 219 million bolívares, while only 187 million bolívares had been collected by the government in taxes from the oil companies. On this evidence Torres was able to bring his memorandum to a close with the following unforgettable conclusion: 'The companies take our oil and the government pays them to take it away.' Gumersindo Torres died in 1946, when I was president. Attended by my military aides I went to the house where there was a vigil over his body, to the great surprise of all the mourners who were there, most of whom were the living remnants of the Gómez régime. I spent two hours beside his coffin, because I wanted to show, in the name of my country, that in Venezuela we do justice to those who have a record of patriotism, honesty and honourability.

The distortion of the economy

Meanwhile the national economy began to be affected by a process of distortion. In 1928 came the first big jump in oil production as 100 million barrels of crude oil were exported. The famous gusher Barroso No. 2 in the state of Zulia had just exploded, which is the only word to use for it. Our traditional exports, like coffee and cocoa, began to decline. The fortunes of the nation and of the government too began to hang by a single thread, that of oil. Some people have argued that the distortion of the economy should be blamed on oil, but a real analysis of the situation gives quite a different picture. If it had been constructively used, the revenue from oil, even the meagre amount which the government collected, would have allowed expenditure on agriculture, stock raising, health and cultural activities to grow at the same pace as the mining industry. In relation to health let it suffice to point out that there was no Ministry of Health. A department of health functioned as part of the Ministry of Agriculture and Livestock, which meant that the same sort of treatment was given to sheep or goats as was given to victims of malaria and bilharzia, endemic diseases which then affected a high percentage of our population. As for education, I will give a single pathetic piece of evidence: in 1930, the only teacher training college in the country produced only one qualified teacher.

The Gómez régime came to an end in 1935 with the natural death of its master. In 1936 a new era began for the country, with General Eleázar López Contreras in power. Many of those who had stayed in the country under Gómez, as well as those who, like me, returned from exile at this time, were well-informed and deeply concerned about the basic problem of the country. I was perhaps the first person to bring out the oil problem in public. I did so on 1 March 1936 at a public meeting, at which I described Venezuela as 'a country, admittedly without a foreign debt, but with its economy dominated by the most daring and aggressive sector of the international financial world, the oil companies. The Venezuelan government may have no foreign creditors, but our country's subsoil has been divided up among the great oil trusts. 80 per cent of our exports are made up of oil, an industry in which Venezuelan capital plays no part at all. And when the government receives 45 per cent of its revenues from that industry, the country is only independent on paper, because in fact it is bound very tightly to the big foreign concerns.' (*El Universal,* Caracas, 2 March 1946.)

The frenzy for concessions continued, and one covering 1 million hectares was granted. Then, early in 1937, the popular parties were banned and their leaders expelled from the country. Some of us went underground to carry on our struggle to 'Venezuelanize' Venezuela and to return to a democratic system with civil liberties.

Infringements of national laws by the companies

I shall go through a series of events in my career in the underground Partido Democrático Nacional (PDN) and its legal successor Acción Democrática. I do not make any implicit criticism of other political groups by doing this. Some groups were not organized as parties, and others adopted different strategy and tactics from ours. The PDN put forward for discussion in Congress a bill to reform the so-called Law of Buoys and Lighthouses, through its single elected deputy, backed by other politically independent deputies who were aligned with the opposition. The Law of Buoys and Lighthouses had been drawn up by the Gómez régime and fixed a tax of 2 bolívares on every ton of oil which left Lake Maracaibo. This was one of the many attempts made by the régime to stop the companies installing their refineries on Aruba and Curaçao. Despite what has sometimes been said, the refineries were in fact deliberately set up outside Venezuela by the oil companies, who realized that the Gómez régime would come to an end and were afraid that it might be followed by a nationalistic government. Therefore they preferred to have the capital for the refinery installations invested outside our country. The companies got round the Gómez law in three ways: first of all, over the carrying capacity of their ships. The ships were not checked at any stage, and so the companies could declare a certain number of tons, without any danger of being proved to be lying, as they certainly were. Secondly there was a clause in the law which stated that, if farm products were exported at the same time as the oil, no tax was due. So the companies put a few bunches of bananas or three or four goats on board and didn't have to pay the tax. Their third way of getting round it was to bring the oil out in small boats because at that time the Maracaibo sand bar was still closed to ships with a draught of any size. They only transhipped the oil into big tankers at Carirubana to take it to the refineries, and so avoided paying the tax.

When we produced our bill there was a great wave of patriotic feeling in the Chamber of Deputies. All the members of the majority party sang the praises of the bill. The President of the Chamber said: 'I can assure the House that this is the most important bill we have had before us as yet.' And it was passed, with modifications which made for a still tighter control of the tax. But as a law it was stillborn and never reached the Senate. Only God and the fixers used by the companies can tell the way in which this law was blocked.

The same sort of thing went on after the political parties had been banned. The companies went on to the attack over legal matters and successfully challenged some of the clauses in laws passed after 1936 which favoured our country and our workers. They demanded what amounted to the annulment of Article 39, which put a partial check on duty-free imports; the Supreme Court decided in favour of the companies against the Venezuelan government. They challenged Paragraph A of Article 27 of the Labour Law which gave some stability of employment to the oil workers. The Supreme Court upheld their challenge and went even further by declaring Paragraphs B and C unconstitutional as well.

The day before yesterday ex-President Caldera recalled the statement by the companies in which they declared that they could not possibly comply with the clause of the Labour Law which determined that at least 75 per cent of the workers they employed should be Venezuelans.

Néstor Luis Pérez

After this came a brief period in which another great Venezuelan following in the footsteps of Gumersindo Torres energetically defended the country's interests in connexion with oil. This was the period in which the Development Ministry was in the hands of Dr Néstor Luis Pérez. He put forward in 1938 a bill which limited the amount of goods which the companies could import free of duty. In this law was mentioned for the first time the idea of setting up mixed companies. It was clearly expressed there, much more clearly than in the law we are discussing in the Senate today, as I shall later show. This law got a stalwart defence from independent deputies in opposition to the government, like Rómulo Gallegos and Doctors Martín Pérez Guevara, Jesús Enrique Losada and Miguel Zúñiga Cisneros.

But it was blocked in much the same way as the Buoys Law modifications were. By sheer trickery, worthy of the most artful characters of all Spanish picaresque literature, Monopolio and Ginés de Pasamonte, the articles which made things difficult for the companies were made to vanish before the law reached the president for his signature. The government had to go back to the official records of Congress in order to reconstruct the text of those articles. The 1938 law got the president's signature a whole year after being passed by Congress and its full powers were never used.

Alberto Adriani

Another nationalist measure introduced in this period was never put into practice either. This measure was the idea of that very promising statesman Alberto Adriani, whose talent, honesty and devotion to public service raised great hopes in our country, only for them to be dashed by his early death. Article 21 of the Customs Tariff Law fixed a maximum duty of 10 per cent on all exports and made special reference to mineral products.

Even in a period when budgets were not balanced, and many of the public's needs weren't met at all, the government was too indecisive to enforce Article 21 of its Customs Tariffs Law. This article became a kind of guilty stain on the nation's conscience – because on every available occasion Acción Democrática supporters upheld it and demanded that it be put into practice.

Steps forward and steps backward

Next came the government of President Medina Angarita. I shall not put on an aggressive or caustic tone of voice to describe this government, nor in other cases either. I shall simply tell the story of what happened to the 1943 bill for the Reform of the Law of Hydrocarbons. This reform was billed as a kind of panacea for the country, and someone even suggested that a public meeting should be held in honour of the man behind the law. I went to the meeting at Los Caobos as the representative of Acción Democrática (of which I was then secretary-general). In my speech I briefly welcomed the auspicious announcement of the government's intention to make legal changes in the relationship between the government and companies and said that we would wait for the bill to be published so as to be able to study it without having made any judgements beforehand and to analyse it and decide what was positive and what was negative in it. The bill was presented to Congress on the very first morning that it met. Both Acción Democrática, through its group in Congress, and the Minoría Unificada of independent deputies and senators, stated their points of disagreement and their points of agreement with the bill. As secretary-general of Acción Democrática I had the opportunity to take a hand in preparing not only the statement with which Dr Juan Pablo Pérez Alfonzo, that great Venezuelan who has so passionately defended our national oil resources, justified his abstention from the vote on the bill, but also the statement with which the Minoría Unificada justified their abstention.

We supported some positive aspects of the bill, like, for example, the unification of the concessions, the increase in taxes on oil, and the clause which forced the companies to keep their accounts inside Venezuela. But we protested that the law in no sense established a just distribution of the income of the producing companies by giving a half share to the government. This was the famous fifty-fifty agreement, 50 per cent for the government and 50 per cent for the companies. We also objected to the argument that it would clear up the industry's problems completely. Some concessions were notorious for having not just flimsy legal bases but totally spurious ones, as was the case with the so-called Valladares concession, which was soon to run out, and from the territory of which the government could have begun to produce oil itself (because it included the San Lorenzo refinery). We also protested when it was said that the law in itself ensured the definitive success of the industry. I only went to secondary school and am not a lawyer, but I know, as any man in the street knows, that laws are organic in nature, not nearly in harmony with reality, but subject to evolution and change.

Between 1944 and 1945 the government gave the companies concessions for exploration and exploitation covering a total of 6½ million hectares, slightly more than the area of all the concessions granted by the governments of Gómez and López Contreras put together.

The events of 18 October 1945

We came to power two years after the 1943 Law of Hydrocarbons had been passed. I shall not now analyse the causes of the revolution of 18 October 1945, a key date, the thirtieth anniversary of which will soon be commemorated by those of us who believe that it produced fundamental changes in the political, social and economic conditions of Venezuela.

I was chosen to preside over the governing Junta Revolucionaria and Dr Pérez Alfonzo, whose abstention over the oil law had received a lot of publicity, became Minister of Development. When we took over the government buildings in Miraflores, instead of celebrating our victorious civilian and military coup with cocktail parties and banquets, we began to sift through the reports and the archives of the Development Ministry. And on 31 December we produced a decree which forced all companies which had made profits of more than 800,000 bolívares to pay a special once-and-for-all tax. Of the 100 million bolívares we received for this tax 88 million Bs came from the oil companies. As a US journalist, Ruth Sheldon, said in *World Petroleum* it was a decree which singled them out. Next we drew up Decree No. 212 which was supported by all the party groups in the National Constituent Assembly. The progressive tax on overall incomes was increased from 9·5 per cent to a maximum of 26 per cent on incomes above 28 million Bs. Only the oil companies reached that figure. There was no talk at that time of applying in Venezuela the courageously nationalistic methods of Lázaro Cárdenas who in 1938 had nationalized Mexico's oil.

In the discussion of Decree No. 212 none of the groups in Congress mentioned nationalization, nor did they mention it when passing the 1948 law which finally confirmed the fifty-fifty relationship between the government and companies. Dr Godofredo Gonzalez, who was a very effective Minister of Development in my government, spoke for Copei, the Social Christian Party; for the URD, Dr Jóvito Villalba, and for the Communist Party, Dr Gustavo Machado. All of them agreed that we had won for the country a just share in the profits of its oil. For Acción Democrática deputy Alberto Carnevali made it clear that the fifty-fifty formula did not represent the limit of the claims for revenue which our country or our party might make. His words were perfectly free of ambiguities: 'If we consider that the commercial situation of the oil industry as a whole shows further improvements, we will not hesitate to go on with the same policy in order to get a still higher level of participation in the oil industry for the nation.' Nobody mentioned nationalization because the situation at the time in Mexico was very different from the situation in our country. The Cárdenas government had nationalized oil in defence against hostilities towards Mexican sovereignty by the oil companies, which refused to obey a decision of the country's Supreme Court on labour conditions. A second, more important reason was that oil was only a supplementary factor in the Mexican economy. Neither at that time nor since then has Mexico been a great exporter of either crude or refined oil. But, we on the other hand, as I have already pointed out and insist on repeating, were hanging by the single thread of oil.

During that period the government drew up the first collective labour contract with the 40,000 oil workers then employed in the industry; the number has since gone down significantly because of automation. I won't trace the steps in the long and difficult conflict which led to the signing of this first collective contract between the companies and their wage-earning and salaried staff in what then amounted to thirty years' history of the industry. That happened on 30 May 1946. *The New York Times* commented on 11 July 1946: 'Under the first collective bargaining contract wage increases alone amounted to 100 million bolívares, which is in itself a triumph. The agreement must be judged a resounding victory for the ruling Junta Revolucionaria under Rómulo Betancourt. A strike in the oilfields would have had disastrous effects on the Venezuelan economy and a decree enforcing compulsory arbitration of the dispute to avoid a strike would have harmed the high reputation enjoyed by the government among the workers.' Beyond that, we set up a commission to study the formation of a national oil company with a national refinery; the commission consisted of the ex-ministers Egaña and Aguerrevere and the much-lamented Alberto Carnevali, who died in his narrow bunk in the prison of San Juan de los Moros under the dictatorship of Pérez Jímenez. We began to realize at that moment that a new problem for our country was already at hand. The oilfields of the Middle East, little known before then, were beginning to be opened up. In these countries, still in a semi-colonial condition, the companies exercised overriding influence, and they were beginning to produce oil that was much cheaper than ours, which was already cheap, and in fact underpriced at $1·50 the barrel. We aimed to send another commission to visit the Middle East countries, but our plan couldn't be carried out because the government of Rómulo Gallegos was overthrown on 24 November 1948, nine months after being voted in by more than a million Venezuelans. I said last night, as I have said in all my books and speeches, and in my official Messages to Congress, that this violent coup was produced by a conspiracy involving the Minister of Defence and heads of the army's General Staff and cannot be blamed on the nation's armed forces as a whole.

Corrupt politics under the dictatorship (1949–58)

This coup brought in the dictatorship which lasted ten years until 23 January 1958. This régime only partly carried out the idea of a commission to visit the Middle East. A group made the trip and returned to Venezuela and the government continued on its inglorious path. I don't think it is necessary to recall its implacable repression against those who fought for human dignity to be respected and to enjoy public liberties, because memories of that period are quite fresh and are deeply engraved on the minds of all our people. I shall only refer to the régime's oil policies. Even though Congress had previously approved in full the policy of giving no more concessions, the ill-sorted Congress of the régime, the members of which were chosen in tightly controlled elections, decided in 1956 in favour of the sale of 820,000 hectares of the best oil lands of Venezuela. Admittedly the government got a good price: 2,500 million bolívares. But the money, like all government revenues, was almost completely sucked up by the chaotic and corrupt administrative practices which in the end brought down the régime, thanks to the Venezuelan people, its energy and its passion for freedom, and also the positive rôle played by the armed forces. This immoral régime had also begun to set up the national petrochemicals company, on the proceeds of its lucrative corruption.

Mosadegh

During this decade occurred an event of great significance for the world oil situation. Mohammed Mosadegh, a descendant of the ancient caliphs and an incorrigible nationalist, managed to inspire his people with a simple, persuasive slogan: 'Persia for the Persians'. He became the leader of a country which had even abandoned its ancient name of Persia to call itself Iran. His Persia was no longer the Persia of the thousand and one nights of Scheherazade, nor the Persia of camels crossing the desert in search of relief in the oases and scimitars at the waist. Persia had become one of the leading oil companies of the Middle East. Its oil was produced under the worst possible terms for the country, by the Anglo-Persian Company, which was, you might say, a direct and legitimate scion of the British crown, which held the majority of its shares, Mosadegh got a majority in Parliament, which then decreed the nationalization of Persian oil.

Alone, pathetically alone, Mosadegh braced himself for every conceivable disaster. But Britain didn't invade his country as it had done in 1946, nor did the Soviet Union, even though the Red Army came over the Caucasus to the very border of Persia. The United Nations Security Council was already in existence, and all the great powers feared the outbreak of a third world war; but a boycott of Persian oil was decreed. Mosadegh went to the USA with his son, who was also his doctor. He was incredibly effective at Lake Success, where the UN then met, with his forceful speeches interrupted by fainting fits, which some considered to be strategically timed, and others simply as evidence that his health was broken. Churchill's government brought Persia before the International Court at The Hague for seizing British property, so Mosadegh went to Holland to defend his country's case – and was successful. The highest court of all for appeals under international law decided that the British claim lay outside its jurisdiction.

But Mosadegh had to escalate a repressive campaign in his own country. The Shah was thrown out and Congress and later the Supreme Court were closed, and finally Mosadegh was overthrown. He had said: 'I shall sleep in Congress because the British would like to see me dead.' But it wasn't just the British who got rid of him. There was a combined operation of the British and American secret services to plan the *coup d'état* which removed Mosadegh. Richard O'Connor, in his very well known book *The Oil Barons*, gives a detailed account of how Mosadegh was overthrown by the secret services of those countries in an operation directed by one of Theodore Roosevelt's grandsons, 'who carried out a real James Bond operation'.

These events in Iran increased the chances for a possible defensive agreement between Venezuela and the oil countries of the Middle East. On this subject Pérez Alfonzo and I exchanged correspondence between the countries in which we were exiled, respectively Mexico and Costa Rica.

Western hemisphere preference and OPEC

In 1959 I returned to power, not catapulted this time by a revolution but voted in by 50 per cent of the electorate in an election I fought against two distinguished Venezuelans, Dr Rafael Caldera and Rear-Admiral Wolfgang Larrazabal. It was my government's policy, as soon as it took office, to pursue further talks with the Arabs and with the North African countries which by then were also producing oil, with a view to setting up not what might be called a cartel but an agreement between producing countries to turn against the cartel of the oil companies, which were closely identified with the industrialized countries of the world and shared out the fantastic spoils with those countries. In 1960, very shortly after I returned to power, President Eisenhower took a decision which discriminated against Venezuela and, theoretically speaking, in favour of Mexico (which actually only exported a small amount of oil to the USA), and also, in more practical terms, in favour of Canada, which does export large quantities to its neighbour. The Eisenhower government fixed a quota for Venezuelan oil exports to the US. My government and also those of Presidents Leoni and Caldera made obstinate representation – demands, I might almost say, to put it in Venezuelan terms – for equal preference to be given to the whole Western hemisphere.

The opportunity to form links with the Arabs arrived very soon. The Venezuelan government was invited to a Pan-Arabian Congress in Cairo. It was a congress to which not only the producing countries, but also the oil companies, were invited. With Dr Pérez Alfonzo providing an unexpectedly lavish budget, Venezuela was represented first of all by the Minister of Mines and Hydrocarbons and by the Director of Cordiplan, Dr Manuel Pérez Guerrero, a Venezuelan of outstanding abilities, who has gone very far as a United Nations administrator, right up to one step below the level of secretary-general. He is proficient in oil affairs and speaks many languages, above all Arabic. Also in Venezuela's delegation in Cairo were representatives of Copei, of the URD, and of Acción Democrática, which were then members of the governing coalition, plus some experts and even some journalists. When the oil companies' representatives were off their guard, there was what looked like a harmless party in the El Maadi Sailing Club in Cairo. Present were Pérez Alfonzo, Pérez Guerrero, Sheikh Tariki, Oil Minister of Saudi Arabia, Salman, the Iraqi Oil Minister, a representative of Kuwait, and one from Iran, closely linked to the Shah, and finally Nessin, Director of the Oil Corporation of the United Arab Republic. Although Pérez Alfonzo and Sheikh Tariki wanted to go further, all that was achieved at the Sailing Club meeting was the signing of a top secret agreement, with a copy for each of the countries represented there. They agreed to set up channels for the exchange of information about oil and to press not for a rise in oil prices but for measures to stop it falling, given that the oil companies had been able unilaterally to fix the reference price on which taxation levels were based, in the Persian Gulf and in the Gulf of Mexico.

In 1960 OPEC was founded in Baghdad. This is an event of enormous historical importance. For the first time in modern history, a group of countries without great battleships or air squadrons, or other kinds of weapons, without even large monetary reserves, formed an alliance against the great powers of the West and against the great oil powers. OPEC grew quite slowly until the special meeting at Tehran in 1971, when it took on its supremely important rôle. But before I go into this, I must refer briefly to the attempts made both by President Leoni and by President Caldera to carry on the policy of defending the nation's interests by stopping over-exploitation of our main source of wealth. President Leoni maintained a policy of firm and decisive support for OPEC, and he got for the government back payments of 800 million bolívares. In the meeting of Presidents of American States in Punta del Este, in discussions with President Johnson of the United States, he stoutly defended the principle that there should be equal preference throughout the Western hemisphere. In his speech here, President Caldera gave a perfectly accurate and detailed account of the efforts made by his government to prepare the ground for the transcendental step we are now taking, in order that the Venezuelan state might get full control over the oil industry.

One of the issues which cropped up during his period in office was the controversy about service contracts. The legality of these contracts had been implied in recent legislation; it should be remembered that even the Constitution of 1961, which was drawn up when all groups, both political and non-political, were opposed to the very idea of concessions, did not exclude the possibility of further concessions, and nor did the 1967 Law of Hydrocarbons. But no one has as yet been able to make use of these loopholes in the constitution or in the law, because it is deeply engraved on the conscience of our people that we have to put an end once and for all to the orgy of concessions which gave away the country's subsoil.

The law, then, allowed for the possibility that service contracts might be signed. There was some opposition to this taking place from the rank and file of Acción Democrática, and President Caldera sent Dr Andres Aguilar, who had been my Minister of Justice until near the end of my term of office, to me as his representative to try to mediate; and so the service contracts were signed. But this story just goes to show that what technical experts like geologists, geophysicists and chemists say should be subject to some scepticism until thoroughly tested. It so happened that the service contracts related to southern Lake Maracaibo, where for a long time the experts had said there was a 'vein of gold'; oil, they said, was easily accessible, just under the water. Now the companies have invested nearly 300 million bolívares, in addition to the 90 million Bs paid in bonds to the Corporación Venezolana del Petróleo, and only Occidental has succeeded in finding oil; and they are still deciding whether it's worth exploiting their well several thousand feet below the water.

The next move was the unilateral fixing of reference prices. This was proposed in a motion put forward by Acción Democrática through their spokesman Arturo Hérnandez Grisanti, backed by the Copei group and the other groups in Congress. The Movimiento Electoral del Pueblo put forward a bill to protect the property of the companies which was liable to revert to the state, and it was passed and became law. During the last government's period in office there was a dramatic change in the international oil situation. In 1970 both the Eastern countries and the West, on which we are better informed (even though there are also some reports on the Soviet Union and the Socialist countries), suffered an acute shortage of oil. This was due to the sharp upturn in the volume of industrial production in Europe and in the USA, and, in addition to that, to the wasteful use of oil, as a cheap multi-purpose fuel. The fast roads of the USA and Europe were thronged by millions of cars going at 70 m.p.h. The situation was also made worse by the Arab-Israeli Six-Day War and the closure of the Suez Canal.

In 1971 took place the historic Tehran meeting, at which the oil countries came to an agreement with the companies which allowed them to do what Venezuela had already done unilaterally, by legislation; namely, to fix the reference price on which taxes raise the level of taxation – admittedly at a rate 5 per cent lower than Venezuela's. The Shah played a very decisive part in the agreement. André Malraux, in his book *Fallen Oaks*, published in 1971, says that at that crucial moment the Persian head of state must have remembered a piece of advice given him by General de Gaulle: 'Sir, you will be offered many ingenious compromises. Never accept them. Devote all your energies to maintaining your independence.' I think that it wasn't so much de Gaulle's advice that was at the back of the Shah's mind as the memory of Mohammed Mosadegh, prophet of the poor countries' defence of their basic wealth. By that time Libya had become a great oil producer. Its government was no longer in the hands of the pliable King Idris, but in those of the highly religious and highly nationalistic Colonel Gaddafi. When negotiating with the companies, the colonel shouted, between verses of the Koran and strong protests against Western imperialism: 'Libya has survived for 5,000 years without oil; it can survive another 5,000 years without it.' This attitude brings to mind the pointed Chinese proverb: 'He who sleeps on the floor does not risk falling out of bed.' Algeria, under President Boumédienne, also came to agreements similar to those of Tehran with the French state companies which exploited its oil.

As for Venezuela, as soon as the government of President Carlos Andres Pérez came into office, it increased the taxation due from the companies which exploit our subsoil. According to the correct figures for oil exports in the first half of 1975, the net income per barrel exported was divided between the state and the companies in the proportion of 95·1 per cent to the state and 4·9 per cent to the private companies.

The time has come for governments to take over total control of the oil industry

We have now reached, after some delay, the right moment for the state to take full control over the exploitation and marketing of hydrocarbons. I say after a delay because this step has already been taken by all the North African countries and those of the Middle East, in Iran, Iraq, the sultanates of the Gulf of Pirates, in the Persian Gulf, etc. Some people object by asking: why should the state take control over the production and marketing of oil, when it gets 90 per cent of its revenues from the industry, which accounts for 40 per cent of the gross national product, and in addition provides 95 per cent of our foreign exchange? I shall have to refute these arguments, which are put forward in all good faith by people who cannot be suspected of being in collusion with the multinational companies.

I have three basic arguments, the first of which is simply patriotism. If a country lets foreign interests exploit its basic raw materials for an indefinite length of time instead of exploiting them itself, it ends up with a distorted, submissive and humiliated colonial mentality. My second argument is a historical one. On this theme I said last night that, in view of the French and American Revolutions, the founding fathers of our republic couldn't send deputies to the Spanish junta at Cadiz in 1809; instead they were fighting for the independence and political sovereignty of our nation. My third argument is that it will be possible to get more revenue from the oil industry under state control than we have up to now in taxes and other payments. This is no superficial argument. In 1946 we decided to take 10 per cent of our oil royalties in kind and use it in trade deals, or sell it in the open market; and we discovered that we got a much better price than the companies had allowed for. In other words, the companies' network of refineries, storage deposits and petrol station outlets is so complex that at all of those stages the companies get hidden profits.

Now, how are we going in practice to make state control over the oil industry work? We can't hope to make it work if we don't realize that myths about self-sufficiency and autarky reminiscent of Robinson Crusoe are quite out of place today. We live in a closely interconnected world, in which nobody can hope to take purely national decisions; so nationalism is not incompatible with internationalism. Now all the countries in the world send representatives not only to the United Nations but also to the Club of Rome to discuss with a global frame of reference the problems that confront mankind. It is obviously in these terms that the governments which have nationalized their oil have thought.

In the Middle East and in North Africa, various methods have been used. In all these countries the state has taken control of the industry, but they haven't hesitated to make specific agreements with the companies which were operating at the time and with the companies which have set up installations since. In Iran the National Iranian Oil Company controls the industry, but it has made agreements with other companies for technical aid.

In these agreements the Iranian state company has kept 50 per cent of the shares for itself; it has even set up mixed-enterprise companies. Iraq is an even more important example because it is the Middle East country with the most radical left-wing government ideology. It is also the country with the closest links, both political and technological, with the Soviet Union. Even though oil production is nationalized in Iraq, the Baras Petroleum Company (BPC) still operates its concessions in the south of the country. In Iraq oil exploitation is very easy because it has a mere 100 highly prolific wells, and there are no possibilities of extending production.

In Algeria the government under President Boumédienne has come to terms with the same French companies which used to produce oil there via the Sunatrach; the French companies get 49 per cent of the shares and the Algerian government keeps the remaining 51 per cent.

I could produce further evidence along these lines, but the details might swamp you. I would like to make what is almost a casual observation. All of you have travelled in Europe, and know that in all European countries the retailing of oil and petroleum by-products is controlled by the state. Even so, in Italy, in France, West Germany, the Scandinavian countries and in my beloved Switzerland, you find the symbols of the multinational companies all over the petrol stations, for example Esso's 'Put a tiger in your tank', or the unmistakable symbol of Shell. I even found the Anglo-Dutch company's shell on the petrol pumps of Budapest, the only capital city in the Communist world which I have visited.

I am in favour of Article 5

Now I shall deal with the issue which has been the central focus of these debates, that is Article 5 of the Organic Law which reserves for the state the production and marketing of hydrocarbons. There has been a real consensus of opinions in Congress and, I venture to say, in the country as a whole, in favour of the state's take-over of the production and marketing of oil. But controversy has arisen in Congress over this

Article 5 of the bill. I shall declare first of all that I fully support Article 5, which allows only two possible loopholes. The first is the possibility of operating contracts granted by the mother company which is going to administer the whole industry. The second involves contracts in partnership with private companies, which the Executive could not put into effect without the approval of Congress granted by a special joint session of the Senate and the Chamber of Deputies. Given that Article 5 does not mention mixed enterprise at all, this possibility of partnerships plays a similar rôle to the safety valves which were incorporated in the Constitution of 1961 and in the 1967 Law of Hydrocarbons to avoid hamstringing the state. A situation could arise in which a partnership agreement could be favourable or even necessary to the national interest. I cannot believe that such an agreement might open another period of submissive surrender of the nation's wealth, because I have faith in Venezuela and in the Venezuelan people, and I know that there will be no more dictatorships in Venezuela, and only dictatorships and dictators can afford not to respect the nation's interests, whether for financial reasons or for any other reason. I am certain that all the political parties, beginning with Acción Democrática, backed by all the parties represented in Congress, will reject any proposal which goes against the basic interests of Venezuela.

These partnership contracts are not on the government's agenda. I am not divulging any dark secrets if I say that highly qualified study groups from the Ministry of Mines have had discussions with the financial groups which exploit our subsoil. They have discussed and reached preliminary agreement about the amount of compensation to be paid to the companies, compensation which has to be paid because there is no question of confiscating their property. Our Constitution states that every investment that is expropriated must be repaid at full value and in a reasonable form. We have practically reached agreement that the price we will pay is the book value of the companies' installations. We have completely rejected the idea of paying what the lawyers call *lucrum cesans*, that is the estimated value of the profits the companies would have made if the concessions had been allowed to run to the end of the terms in the contracts, i.e. until 1983 and 1984. There has been no discussion of partnership contracts, simply of technical cooperation for three years under contracts which may be renewed in the future. Meanwhile the present government, fulfilling the hopes of all of Venezuela, has set aside 20 hectares next to Simon Bolívar University as a site for the Instituto Venezolano del Petróleo.

We have our own Venezuelan experts and also foreign experts who will transfer from private enterprise to the state companies. But we must train future generations to take over from them. We must train people to understand the sophisticated technology of petroleum. We have sent a lot of students to the United States and to Europe, but we must also educate them here in Venezuela.

The government has taken care to choose as directors of the future mother company of the nationalized industry (which is to be called Petróleos de Venezuela, S.A.) citizens with excellent records of technical ability and honesty in administration and elsewhere. Here are their names: General Rafael Alfonzo Ravard, who was president of the Corporación Venezolana de Guayana for fifteen years and has now been replaced by one of his protégés, another honest and technically very knowledgeable administrator, Dr Argenis Gamboa. Alongside General Ravard are Dr Julio Sosa Rodriguez, of whom, two days ago, ex-President Rafael Caldera made a justifiable and emotional eulogy, Dr Carlos Guillermo Rangel, Dr Julio Cesar Arreaza, Dr Benito Raul Losada, the deputy and union leader Manuel Penalver, to represent the workers, and Doctors Edgar Leal, Domingo Casanova, and Alirio Parra. They will form the board of directors of PETROVEN, S.A.

In the companies which will depend on this mother company we will place honest and capable administrators, in order to avoid two dangers of which we are forewarned as Venezuela takes absolute control over its oil, those of too much bureaucracy and corrupt administration.

Finally, honourable Senators, I shall give you one important fact for comparative purposes. The Soviet Union, which has made great strides in technology, as is well known, and is now challenging the USA for economic and military supremacy in the world and in space, came to an agreement in Paris in December 1974 with representatives of Japan and the USA (as was reported not only in specialist publications like *Petroleum Intelligence Weekly* and *The Petroleum Economist* but also in the international press) to form a partnership to explore and subsequently exploit the gas reserves of Siberia. What is more, the Soviet Union, having reached the right stage for pragmatism, has signed partnership agreements with one of the most aggressive oil companies in the modern world, Occidental Petroleum, controlled by the audacious Dr Armand Harmer. So, if a great power like the Soviet Union is doing this, why should we be worried or afraid about discussing with some companies, with full knowledge of what we possess, when to build up stocks, when to renew our pretty obsolete refinery installations, and also how to begin to explore, not to exploit, the famous bitumen band in the Orinoco River. Ex-President Caldera's speech was very informative about this. He said that during his term of office more than 100 million bolívares were spent on explorations in that area. It has a special kind of oil, totally unknown in the world, which consists of pitches mixed with sand, a lot of sulphur and varying quantities of metals. The bitumen band is said to contain reserves amounting to thousands of millions of barrels. Then why not begin to explore it further, but not in order to exploit it? This government has started a policy of conservation of resources, involving reducing the production of traditional oils. Why should it be suspected of wanting to exploit the Orinoco bitumen band? In any case, the experts, if they are agreed on anything, agree that the area could not be exploited in the next ten years. By then Carlos Andres Pérez will have left office and there will be another President of Venezuela.

This is in fact the point I wanted to make and I don't want to put further emphasis on this subject. What I do wish to make clear, speaking not only as an active member of my party, but also as a Venezuelan who follows the dictates of his conscience, is that I think the much discussed Article 5 should become law along with the rest of this bill.

Mr President of the Senate, distinguished ex-President of the Republic, life Senator Dr Rafael Caldera, honourable Senators, and members of the public, I shall end with an appeal which may be thought naïve. I call for this bill to be debated with less bitterness in the Senate than in the Chamber of Deputies. I have seen copies of all the speeches made there. Their language sometimes became very violent. It reminded me of an essay I read a long time ago by the great Spanish writer and philosopher José Ortega y Gasset, called *Chinos y Chinitos.* It tells how, when Ortega y Gasset was passing through Peking and other Chinese towns, then under the rule of the feudal Emperor, but now under Mao and Chou-en Lai, his attention was caught by Chinese sitting on the roofs of their houses, gesticulating and shouting what he took to be obscenities. He asked his interpreter what they were doing and the interpreter told him: 'We Chinese are so polite that sometimes we get exasperated. When you play chess, you have to say to your opponent, "Allow me to take your honourable pawn with my miserable queen." But there comes a time when we can't bear the tension that builds up as a result of this kind of politeness, and we get on to the roofs of our houses and shout the worst insults we can think of to "clean out our chimneys".' Now the deputies of all the different parties have already cleaned out our chimneys. I hope that my naïve proposal, which I make in all good faith, will be accepted and that the debate in the Senate, which is after all the chamber of elders, is a calm one.

I shall end by saying that I have absolute faith in the success of the government's take-over of control after the law nationalizing the production and marketing of hydrocarbons has been passed. The government which will begin this vital task for the country is headed by Carlos Andres Pérez, a brilliant and skilful statesman, who, after getting the highest number of votes achieved by any presidential candidate in the whole democratic history of our country, has taken some daring decisions and has also shown a capacity to realize his mistakes and correct them. The mother company which will administer the nationalized industry (Petróleos de Venezuela, S.A.) will be controlled by experts with great technical ability and administrators of unblemished honesty, and so will any other companies which we may need to set up. The Venezuelan and non-Venezuelan technicians who will continue to be employed in the industry have a considerable accumulated experience. The oil workers have sufficient political awareness to serve the national company with the same devotion as they have served the multinationals up to now. All Venezuelans, including those who have exercised their democratic right to disagree with specific parts of the law which we are about to pass, will make a nationalistic contribution to the success of the new administration of our basic wealth in the hands of the nation.

We are about to take a historic step of great significance. The men and women of this country will act decisively, as we have done on all the other occasions when Venezuela has needed us, with a keen awareness of our responsibility at this particularly important moment for the country.

Caracas, 6 August 1975

Oil: Venezuela and the World

(I wrote this short essay in Berne in December 1967. It covers the politics of oil from 1958 up till that date; in other words, the policies of the Junta de Gobierno which took over when Pérez Jiménez had been overthrown and of the constitutional governments over which I and then Dr Raul Leoni presided. It is not for any uncharitably sectarian political reason, but simply because of the date when it was written, that I do not consider the oil policies of Dr Rafael Caldera, founder and leader of Copei, the Social Christian party, during his five years as president. He carried out a policy of consistent nationalism, including concrete measures in defence of national interests in connexion with the oil question.)

Oil policy during the decade from 1959 to 1969, under the governments Rómulo Betancourt and Dr Raul Leoni

On 19 December 1958 the Junta de Gobierno led by Dr Edgar Sanabria promulgated a decree which put the rate of taxation on the oil industry up from 52 per cent to 65 per cent.

This was almost the last act of the *de facto* government, the days of which were then numbered because I had already been elected president. It provoked angry reactions from the international companies which exploited our subsoil. The manager of the Creole Petroleum Corporation (Standard Oil) was less restrained than his colleagues from the other companies in the 'Big Three', Shell and Mene Grande. This was Mr Harold Haight, who attacked it with a threat which the companies were subsequently to carry out. In a press statement at Maiquetia (Caracas) he aggressively criticized the decree, and, not content with dropping hints, coolly predicted that it would produce a fall, not a rise, in the revenue collected by the Venezuelan government. These were the words he used: 'It is extremely doubtful whether by raising the taxes the government will get more revenue than it would have done under the fifty-fifty agreement. Time will tell, but time normally tells a little late.'

The Junta got support from all sides in this situation. Mr Haight's entry visa was cancelled to prevent him returning, but the threats he had made were carried out. I have already recalled that in my first Message to National Congress I promised that my government would put the Junta's decree into practice.

Some of the finer points of the retaliation by the companies for the decree were not fully understood at the time. As recently as 1975 has this been put across in full detail and with accurate figures by an American professor of undoubtable prestige in his country, namely Franklin Tugwell, Associate Professor of Politics in Pomona College, in his revealing book, *The Politics of Oil in Venezuela*, published by Standford University Press, in California.

In 1959 the companies began a well coordinated plan of disinvestment in Venezuela, and at the same time they intensified their explorations in Canada and in the Middle East. Tugwell says, on p. 77: 'To carry out their threats, the most powerful companies embarked on a programme of gradual disinvestment in the Venezuelan oil industry, and at the same time began to increase production in other areas under their control (especially Canada and the Middle East). Investment, exploration and drilling activity declined, and so did even the numbers employed, which went down by 28 per cent between 1960 and 1966.' Tugwell also shows, on the basis of the statistical tables in the annual reports of the Creole Petroleum Corporation 1972–3, that in comparison with 598 exploratory wells in 1958, there were only 100 left in 1968.

The other device in the long-term strategy used by the companies to force the Venezuelan government to lower the taxes on oil and open tenders for new concessions was the organization called the Cámara del Petroleo which they formed in 1959, and then had affiliated to the association of Venezuelan businessmen in the private sector, FEDECAMARAS. Here Tugwell's arguments and figures imply serious charges, to which that businessmen's organization, normally quite critical of the nationalistic oil policy of post-1958 governments, should reply. Tugwell says (on p. 80): 'The latest payment made by the oil companies to FEDECAMARAS amounted to 400,000 Bs, twice as much as the second biggest contributors. Individual oil companies or their representatives often made special *ad hoc* contributions for specific publicity campaigns.'

The situation of the oil industry in the country was one of the gravest problems which confronted my government when I took office on 13 February 1959.

No longer was there a fierce demand for oil as there had been during and after the closure of the Suez Canal in 1956. The main customers for our oil exports – Western Europe and the United States – were successful in a united effort to get what was in their interest – a reduction of the selling price. An unbroken period of ten years' rising production came to an end, and the amount of oil extracted from Venezuelan wells went down below the 1957 figures. The prosperous years of which the Pérez Jiménez régime would not or could not take advantage had come to an end. The dictator's government had not pressed for more out of the companies because it wanted to ingratiate itself with such powerful interests, and thus with the governments with which they were linked; it did not dare to claim a higher proportion of the surplus profits made by the oil companies because that could only be done by a government which was immune to bribery and certain to have the Venezuelan people behind it not against it.

In addition to these challenges facing the government that took office in 1959 arose the problem of the system of restrictions on oil imports set up by the US government.

As we told General Eisenhower's government firmly and decisively, this policy would directly affect Venezuela, which still sells most of its oil in that country. A meeting of my cabinet approved the text of a statement which we sent to the State Department on 24 April 1959. Two excerpts from that statement will illustrate its highly justified tone of protest:

> The Venezuelan government is particularly worried by the implications of the United States government's policy of discriminating against Venezuelan oil on the pretext of the national security of the United States. This policy implies that Venezuela, and her oil resources, are no longer considered vital to US security, as they have been up to now. . . . The Venezuelan government will be put in a very difficult position by any measure under the Mandatory Programme for Oil Imports which discriminates in favour of one or two countries for security reasons and excludes the other countries of the Western hemisphere.
>
> Any privileges granted, whether explicitly or implicitly, to oil imports from Canada and Mexico, without an immediate guarantee of similar privileges for Venezuela, will have an adverse effect on Venezuelan public opinion and will certainly contribute to increase hostile feeling against the USA.

Exactly what we had feared and foreseen happened on 30 April 1959. A message from the US President excluded from the import restrictions crude oil and its by-products which arrived in the US 'by pipeline or in road or rail tankers from the country of production'; in other words, oil from Mexico, which exported very little to the US, and from Canada, a big supplier of the US. (I am convinced that the policy of restricting imports of Venezuelan crude oil to the US, and I will say it in brutally frank terms, was part of the strategy of the powerful oil companies to force my government to go back on its nationalistic policies. This is hardly an unsupported allegation, for it is well known that the

Eisenhower administration was, among recent US administrations, one of those which was most heavily under the influence of the oil companies.)

The note sent by the Venezuelan government to the State Department let the US government know that we would make a stand and not be weak-kneed about this plainly unfair decision against the Venezuelan economy. President Eisenhower sent me a letter in which he attempted to justify his government's policy with the unconvincing argument that oil which was imported 'by road, pipeline, or rail' from neighbouring countries needed special guarantees to ensure the continuity of the supply in times of emergency. He referred to a statement made by a White House spokesman on 10 March 1959, and added, pointing out that the US and the other oil-producing countries of the Western hemisphere had the same interests 'in the wider context of the security of the free world': 'Given this fact, talks will continue with Venezuela and other Western hemisphere countries with a view to finding a coordinated solution to the oil problem, which is closely connected to the national defence and other national interests of the producing countries.'

In the reply we sent to this letter from the US President the Venezuelan government repeated and strengthened its protests against the damage being caused to our economy. In addition to this we decided that, if the US could decide unilaterally to limit our exports to that country, we too could take unilateral steps to restrict their trade in the opposite direction. This is how our position was correctly stated, by the Minister of Mines and Hydrocarbons, Dr Jose Antonio Mayoe, in National Congress, on 31 May 1967:

The very day that the US imposed its restrictions, Venezuela retaliated, as President Betancourt announced in a speech on that occasion, by fixing import quotas for the majority of the articles on List No. 1 of the Treaty between the United States and Venezuela. From then on these products began to be manufactured in this country. As Betancourt's government clearly said more than once, for us national security means economic development, so we are simply putting into practice the clause about our national security.

The Venezuelan government kept on pushing the US hard with this argument. We accepted that the privileges granted to Mexico and Canada were fair, and simply asked, with solid and unshakeable arguments, for the same privileges to be extended to Venezuela. Washington recognized that our demands were reasonable, but gave as an excuse for not agreeing to them at once the argument that they had to protect the interests of the independent oil producers inside the US who were only allowed by government regulations to extract oil in very limited quantities, which meant that they could work their wells for not more than a few days each month.

When Kennedy came to the White House we were still fighting for justice for Venezuela. Even before he took his presidential oath, Kennedy sent several emissaries to Caracas, some of whom were personal friends of mine, whom I had got to know while I was in exile in the USA. We had long and detailed talks which naturally centred on the trade restrictions. Perez Alfonzo, whom I had brought back to do the job of

Minister of Mines which he had carried out with great efficiency and absolute dedication from 1945 to 1948, justified our stand methodically and with a wealth of evidence in those long talks. In Washington it had been made clear that the new president was ready to make amends for this piece of injustice and resolve other problems which beset trade between North and South America. They suggested the idea that our government should be consulted before any changes in the quotas fixed for imports from Venezuela were made.

The system of consultations which then started provides a revealing story. During 1960 the US government let us know that President Kennedy wanted to install a direct telephone link by which he and I could speak to each other whenever we liked without any delay. I agreed and receivers were installed in my office at Miraflores and in my bedroom at Los Nuñez. Given my irrepressible *criollo* sense of humour, I couldn't resist telling this joke to my friends: 'These telephones are only little sisters of the hot line between the Kremlin and the White House. I should think they'll get mouldy for lack of use.' But I was wrong, for a month later I was using the much-maligned telephone link. We were consulted with only a very few days to spare before Kennedy made a declaration on oil quotas. The declaration did not make the situation worse for Venezuela, but we couldn't take the delay in consulting us lying down. I got in contact with the White House, which put me through to the Florida beach house where Kennedy was spending his Christmas holidays. Beside me I had my daughter Virginia listening in on another telephone to act as interpreter. I was angry and used strong words, but Virginia, when she translated them, took the sting out of them, though I demanded that she should make a literal translation. In any case President Kennedy at once sent off a telegram asking whether the Christmas holidays would get in the way if he sent a personal representative accompanied by an expert from the Department of the Interior, to which all oil matters are assigned in the USA. We replied at once that we welcomed the mission and that the New Year celebrations would make no difference to the talks. Along with my Ministers of Mines and Finance I had several meetings with Kennedy's representatives in which we discussed the oil quota problem. As we defended the national interest aggressively and with reasonable arguments, shut up in my presidential office at Miraflores, we could hear outside the explosion of happiness with which the people of Caracas celebrate Christmas and the New Year. Kennedy was quite conscious of the disagreement between me and my daughter when she took the sting out of my words, and when he came to Venezuela the three of us laughed when he quoted the well-known saying, 'Traduttore, traditore'.

During President Kennedy's visit of a few hours to Caracas, the interview which my cabinet ministers and I had with him were again taken up with the subject of the oil quotas.

The joint declaration signed by President Kennedy and myself in Miraflores on 17 December 1961 mentions implicitly, and also, I might add, all but explicitly, exactly what Venezuela wanted in terms of its oil trade with the US. Paragraph 6 of that declaration made a concrete reference to the quotas and the discrimination against our oil, though it was wrapped up in rather opaque diplomatic language. It says:

> Both presidents declare themselves in agreement that a great effort in the field of social relations, in accordance with the principles of the Alliance for Progress, must be made to go hand in hand with the process of economic development. The price of basic products and the commercial practice of the importing countries should in an effective manner take into account Latin America's dependence on its exports. The realization of this fact is vital to the maintenance of the letter and the spirit of the Charter of Punta del Este.

When I accepted President Kennedy's invitation and made a state visit to the United States, I did not react to the American's superficial attempt to appeal to my, in fact, non-existent vanity by playing the Venezuelan national anthem in my honour as I reached the entrance to the residence of the head of the most powerful government in the world. I went there to continue my tenacious and unflinching attempt to press home the unchanged claim for fair treatment for our oil exports to the US.

President Kennedy and I, accompanied by our advisers, discussed this problem for four hours in the White House. I spoke first and then President Kennedy replied carefully, with figures supplied by his aides. In a whisper I asked my Finance Minister Dr Andres German Otero: 'Have you got sufficient evidence to refute what Kennedy's saying?' He calmly replied: 'I haven't got any figures in writing, but I've got them all in my head.' Phlegmatically, like an Englishman rather than a man of the tropics, Otero capped the President's statistics one after another with figures produced and analysed by our ministries of Mines and Finance. The meeting ended with an agreement to set up a bilateral commission between the two countries. That evening there took place a reception given for us by the Kennedys, at which the President said to me more or less these words: 'I want to congratulate you on the frankness with which you spoke to me, and also your Finance Minister on the precision of his statistics. I have had the figures I used checked and I am convinced that your minister's figures are the correct ones.' I called Otero over to us so that he could hear from our host's mouth this praise of his powers of exposition and his will in handling correct statistics.

As for my 'frankness', I had made no bones about what I had to say. In the interview I had alone with President Kennedy, I asked if it would be possible to dispense with the interpreter for some of the time. In my rather uncertain English, which becomes quite fluent when I really need to make myself understood, I told him that the establishment of oil quotas for Venezuela and not for Mexico or Canada not only was unfair, but also had given rise to racket, a dirty game (those were my exact words). There was a traffic in the permits to import Venezuelan oil granted to small refineries. The *Oil and Gas Journal* had pointed out in an editorial that, 'The aim was to get a wider spread of the profits on low-cost foreign crude oil. The refineries have discovered that an important quota is as valuable as ready money. They can deal in permits as if they were cash, at a rate of 50 cents to one dollar a barrel.' I went on to say that there was no point in having the Alliance for Progress, and it would end up as pure demagoguery, if they could not help to resolve injustices like those suffered by Venezuela; my country needed to make up lost ground on its path towards development and had great material and cultural gaps to deal with, but was being robbed by the tricks of professional pickpockets, to the tune of nearly 1,000 million dollars at that time, in order to enrich a country of many multimillionaires still further. Finally I pointed out that Venezuela would continue to be plundered by the United States so long as Venezuelan oil wasn't treated in the same terms as Canadian oil and so long as there was still a scramble for import permits among a few hundred influential vote-catchers in the US. President Kennedy looked at me with his typical penetrating stare with one eye half-closed as he listened attentively to my halting but passionate protest, and with a quiet humility which indicated inner strength rather than brash self-confidence. His reply was swift and sincere. I can't quote his exact words, but the gist of them was that 'the whole problem was bedevilled by a combination of private interests' and that he was sure that this obviously unfair situation would be resolved in a way favourable to Venezuela before the terms of office of both of us came to an end. So the main reason for my contentment as I returned to Venezuela, and for the great optimism which I tried to convey to the people of Venezuela when I spoke at Maiquetia airport, was that I had this very explicit promise from the President of the United States.

Then came Kennedy's tragic assassination at Dallas, which meant that he was unable to fulfil the promise he had made to me. I left office, but that did not put an end to our sustained efforts to get better treatment. The problem of the oil quotas was again brought up by President Leoni when he spoke to President Johnson at the Punta del Este meeting. As recently as 1967 there was a unanimous wave of national protest when it became known that US Congress might pass a law to ratify the policy of restrictions on oil imports from Venezuela permanently. The situation is still totally unfair and the united efforts of the Venezuelan people and their government must not cease until we are able to put an end to this damaging blight in our commercial relations with the USA.

This stalwart action in defence of the country's rights has never been boasted of in public speeches by governments since 1959. For those who bear the responsibility of running this small country which is rich in the most sought-after raw material in the world today there are two different styles which can be adopted for commercial relations with the great industrial powers. One style uses words taken from the babble of public meetings, and the other uses unappealing but unrelenting daily pressures to make sure justice is done. I would recommend that the hotheads of the left, the noisy extremists who reject this course of action, should read a play by their idol Jean-Paul Sartre, *In the Mesh.* The central figure of this Sartre play is Jean Anguerre, a revolutionary who obstinately defends the oil interests of his little country; he refuses to go to extreme lengths against the oil companies and the heavily-armed governments which support them and would bring disasters on the heads of his people. It is one thing to govern a country with energetically but not unthinkingly nationalistic zeal without making it teeter on the edge of an abyss. It is quite another thing to shout down microphones in Congress or in public demonstrations the slogans of a verbose and irresponsible radicalism.

A reasonable share in the profits of the industry and improvements in wages and social services for the oil workers

In connexion with oil the democratic government was not active only in fighting the US over the quotas. At the same time as our attempts to get concessions in that direction, we hit the companies hard as well.

The Ministry of Mines and Hydrocarbons set up a Coordinating Committee on the Conservation and Marketing of Hydrocarbons. This became very useful to the government as a technical body manned by highly-qualified experts, who had the right to intervene in the industrial and commercial activities of the oil companies by monitoring their techniques for extracting the oil and selling what the wells produced and the by-products manufactured in the refineries.

It was a simple matter for this committee to provide us with evidence which showed that, despite the damage being done to trade with the United States, the government was getting a high percentage of the value of each barrel of oil as revenue, and that the unacceptably high percentage profit which the companies had been making had gone down significantly. In addition, free bargaining by the reorganized trade unions with the companies produced a rise in the oilworkers' incomes, both in terms of wages and in terms of social services, above the levels enjoyed by workers in other parts of the world.

In 1963, during my term of office, the powerful FEDEPETROL (Federation of Oilworkers) signed its first collective contract with the companies. Professor Peter Odell, of the London School of Economics, produced a table (Table 1) to illustrate the amounts by which workers' income rose, as follows:

Table 1

Pre-1963		*Post-1963*
Bs 29·39	Basic daily wage (per worker per day)	Bs 32·70
Bs 29·90	Other cash payments (housing, transport, etc.)	Bs 40·60
Bs 17·20	Cost of services (social services in oilfields and refineries)	Bs 19·55
Bs 9·80	Terminal bonuses	Bs 11·60
Bs 86·26	Average daily wage	Bs 104·45

Professor Odell commented on these figures: 'These show, as the result of the 1963 negotiations, a rise of 20 per cent. Published figures show that the hourly earnings of an oilworker in Venezuela is $3·67, as against $3·50 in the US.'

The collective contract signed under President Leoni in 1966 meant new achievements for the oilworkers. The oilworkers' leader Luis Tovar, one of the oldest and most capable union leaders in the country, won world recognition for his toughness and skill when he was made president of the World Oilworkers' Federation, which has several million members. (He died in 1974.)

This policy meant that the government did not just consent to the unionization of the oilworkers, but actively helped the unions to grow, and the government's approval of the results they achieved in the collective contracts had three important effects. First of all the purchasing power of those who worked in the most prosperous business sector was justly increased, which directly improved the market for agricultural and industrial products. Harmony between workers and management was guaranteed, and with it a continuous flow of oil exports. Finally, infiltration by Communists, who had only had a foothold in some small unions in the oilfields and in the towns in the oil zone, was stamped out.

The increase in government oil revenues can best be appreciated by making a brief survey of some simple figures (despite a fall in the selling price of oil and increased competition because of the new oilfields brought into production in Libya, Nigeria, etc.).

My party colleague and close friend and Dr Raul Leoni's Minister of Mines and Hydrocarbons, Dr Mayobre, was able to point out, in a speech to National Congress on 31 May 1967, the following basic facts (which, since it's difficult to get round the truth, haven't yet been challenged):

> Venezuela's systematic oil legislation assigns a higher proportion of the total value of the oil produced to the government than in any other country. In 1956, as is well known, the state got 52 per cent and the companies 48 per cent. In 1966 the proportion is 65 per cent for the government and 35 per cent for the companies. The amount of tax paid per barrel went up from 3·39 bolívares in 1956 to 4·06 bolívares in 1966.

These figures basically coincide, though there are some differences with the figures given by Pérez Alfonzo in his book *El Pentagono Petrolero* (p. 20) as follows:

	1950–7	*1958–64*
Value of the state's share	Bs 2·64	Bs 3·83
Proportions taken by		
a) the companies	a) 42%	a) 27%
b) the state	b) 55%	b) 73%

Pérez Alfonzo brings out the significance of these figures with a careful commentary: 'The progress we have made is obvious. Despite falling prices, the government gets 46 per cent more per barrel than it got in the 1950–7 period, and the companies get 35 per cent less per barrel. This makes the ratio between the companies and the government 27–73 as against 42–55 in the earlier period.' Pérez Alfonzo makes it plain that he believes that this system of payments would be more equitable and more beneficial to our country if the figures for the companies' profits were calculated in relation to the net value of their fixed capital. He argues therefore for a selective tax which he justifies in this way: 'The companies which are lucky enough to have the wells which produce most are those which will have to pay more, and will thus help to bring about a further increase in the state's share in the industry's profits.'

This coherent and reasonable suggestion deserves further attention because it is based on the correct presumption that the productivity of a given area depends more on the chance disposition of suitable oil deposits than on skilful, efficient administration by the drilling companies. These arguments should definitely hold the attention of those who govern Venezuela. For oil taxation policies should be dynamic, not static, and should try to explore all the possible routes by which the country could get more benefits from its privileged position as one of the world's leading producers of this much-needed source of energy. Even so, our democratic governments between 1958 and 1967 have managed to put Venezuela in the position of getting more revenue from oil than is the case in any other country in the capitalist world of the market economy.

At the same time as maintaining this peculiarly Venezuelan policy we fully recognize that the companies have a right to their legitimate profits.

Successive occupants of the government offices in Miraflores weren't mad. The companies, which after all are capitalist enterprises and not religious foundations dedicated to charity, could have given us a very difficult time if we had driven them into a corner. We have never worshipped at the shrine of demagoguery. A severe recession affected the country (and the government's revenues). Without stopping to think, the provisional government at once paid off the large and highly questionable floating debt left by the dictatorship; that cost over 1,000 million bolívares and left the National Treasury almost empty. The ill-conceived Emergency Plan, which, though intended to mitigate the effects of unemployment at that time, amounted to paying people not to go to work, cost another 600 million bolívares. My government made the mistake of increasing expenditure in 1960–1 when the meagre treasury reserves couldn't cover the deficit produced by such a high level of expenditure. Acción Democrática stuck, and still does stick, so closely to its programme that even in this period of serious economic and fiscal difficulties we never thought of taking an easy way out and putting up oil concessions for sale. Instead we held rigidly to our policy of granting no further concessions, as I shall show later on, and we did not put up the oil taxes. We preferred to make the government feel the pinch by cutting expenditure, without stopping services or freezing investments, and to appeal to the people, from whom we got a magnificent response, to pay more in taxes and spend less on luxury goods. This was what our dear, or perhaps not so dear, friends the Communists nicknamed the 'hunger law'.

This was the moment at which the government made it plain that it did not want to drive the oil companies at work in the country into a corner. We defied the loud protests of the Communists, who were then glorying in the success of Fidel Castro, and of others who echoed their words, and made it clear that at that moment taxes on oil would not be increased because the whole industry was making only modest profits. I was very explicit when I read a special message to National Congress on 4 May 1961, and said: 'We have said that the present profits of the oil companies, which average 11 per cent, are low because they aren't high enough to go on attracting foreign capital which could get the same rate of return inside its country of origin. We are not prepared to reduce our government's share in the profits, because we maintain that the present low returns in the industry stem from the low price of oil, and that the only hope for higher returns lies in trying to get a rapid rise in prices.' The Minister of Mines and Hydrocarbons, Pérez Alfonzo, spent a whole day answering questions in Congress on this subject, defending our position with detailed facts and figures. It was quite true that at that moment the profits of the oil companies had gone down sharply to 12·87 per cent and 13·12 per cent in 1960. We realized that the profits would go up in the future, but that at that very time the figures were the wrong side of the danger line. If we had put the taxes up, it would have meant simply encouraging the companies to shut off their Venezuelan wells and open up others in the Middle East. Even though we presented a very solid case to Congress, and had the backing of the parliamentary groups of both Acción Democrática and Copei, Congress made the oil companies liable to a new tax on profits. But the tax was designated optional, not mandatory, for the Executive and we chose not to make use of it at all, because Congress had prevented us from making an exception for the oil companies, and we were sure that it was the wrong moment to put up the taxes on them. In the end we were proved right. The profits of the oil companies began to go up and it became therefore both just and feasible for Leoni's government to increase the rate of income tax, to reach an agreement on compensation payments and to set up a system of fixing reference prices by which we could ensure a definite, stable price for our oil.

The policy of 'no more concessions' and the formation of the Corporación Venezolana de Petróleo and of OPEC

In its first period in office, between 1945 and 1948, Acción Democrática took a firm stand in defence of our national oil interests. In one chapter of my book *Venezuela, Política y Petróleo,* I explain what the government did in those three years, and point out how, even in such a short period, the government laid the foundations of a coherent oil policy which had a definite goal, namely the nationalization of Venezuela's basic industry.

One of the main tenets of this policy can be summed up in a short but expressive phrase: 'No more concessions'. I have already shown how a huge chunk of national territory had been conceded to foreign interests, under a system which virtually turned us into a colony again.

We had very different ideas from those submissive incompetent governments which cared not at all for the national interests. Given that the industry had been going for fifty years in our country, there was no justification for the fact that the state had only very marginal contacts with it, limited to tax collection. It was of the utmost importance that we should organize a Venezuelan state company, as had been done elsewhere in Latin America and in the Middle East.

This company should take an active part in the oil business, exploring for oil and producing and marketing it. The company should be assigned free zones, above all in areas where oil had already been found, which it could then work either directly or in partnership with private companies. But it would not be handcuffed any longer by the system of concessions; instead it would work under the more flexible system of 'service contracts' which would guarantee better returns for our country.

To carry out this policy I set up, by presidential decree, on 19 April 1960, the Corporacion Venezolana del Petróleo (known as the CVP). We deliberately chose the same day as the anniversary of our declaration of political independence, because we wanted to show that we attached special importance to this step in our struggle to conquer Venezuela's economic independence.

We took care, while setting up the CVP, to prevent it from becoming another overstaffed bureaucracy. We never wanted it to make an ostentatious attempt to compete with the gigantic private companies. During the seven years it has been in existence the CVP has successfully drilled many highly productive wells. Its sales cover 16 per cent of the Venezuelan market for petrol, which it is trying hard to push up to the 33 per cent which it was assigned by a government decree. It supplies half the gas consumed in the country, in the Caracas area and in the surrounding districts as far as Barquisimeto, and next year it plans to bring gas from Lake Maracaibo to El Tablazo and from Anaco to Puerto Ordaz. It also enlarged and now owns the only state oil refinery, at Moron. In its oilfields at Boscán, Oriente and Barinas the CVP has reached for a start the figure of 20,000 barrels of oil a day, which will go up to 70,000 barrels a day in 1968. Its capital is just under 400 million bolívares, and, in seven years, it has contributed 72 million bolívares to the state in terms of royalties, income tax, and net profits. Most of the private oil companies were reluctant at first to cooperate with the CVP. But they have begun to accept the gradual growth of the state company as something which cannot be revered because it fulfils one of the deepest desires of the Venezuelan people, and have therefore changed their attitude. Significantly, in mid-1967, fourteen of the private companies signed an agreement with the CVP for joint geophysical explorations in the Gulf of Venezuela and are now negotiating an agreement by which they will get hold of the geophysical data obtained by the CVP in the southern zone of Lake Maracaibo.

Another basic tenet of our national oil policy going back to 1959 has been our membership, alongside the Middle East countries, of the Organization of Petrol-Exporting Countries (OPEC). The Baghdad agreement which was the starting point of OPEC was made public on 24 September 1960. The founder members were Venezuela, Iraq, Kuwait, and Saudi Arabia. Subsequently the other oil-producing countries joined the organization.

Dr Juan Pablo Pérez Alfonzo is known throughout the world as the 'father' of OPEC. He has always insisted that his influential position in oil matters has been won thanks to Acción Democrática and above all to Rómulo Betancourt. In his book *Hundiendonos en el Excremento del Diablo* (Caracas, Editorial Lisbona, 1976, p. 173) he writes:

> This world-shaking series of events which has raised hopes of a new economic order would not

> have come about without Acción Democrática, and in particular without the vision and constant dedication of its leader Rómulo Betancourt. . . . I feel that it's necessary to repeat once again that none of my achievements in these vital oil matters would have been possible without the stimulus of Betancourt and the help of his party. Above all at the very beginning I would not have started at all had it not been for this active politician's promptings which forced me to commit myself to action. I wouldn't have written the justification of my 1943 Abstention without the help of Betancourt, who even typed out for me what I wanted to say. I wouldn't have joined his government if he hadn't had me virtually dragged out of my house here at Los Chorros by my neighbour Ricardo Montilla. And I certainly would not have returned from Mexico to shut myself up in the Torre Norte del Silencio if Betancourt hadn't persuaded me, in December 1958, soon after he had been elected president, that we had to take our oil policy further. What we had started in 1948 had been cut short by the military coup.

OPEC is a united front put up by the oil countries so as to maintain a high price for the oil drilled inside their boundaries by the international oil cartel. Its second aim is to coordinate the efforts made by the governments of the member countries to get the highest possible share in the profits made from their mineral wealth. This organization fits in with the strategy employed by the developing countries of uniting to defend their rights which are either haggled over or ignored by the great industrial powers and the monopolistic multinational companies. Like all the recently founded organizations of this kind, OPEC has had some very difficult and trying obstacles to overcome. The big oil companies never viewed its formation kindly, and treated it in fact with scarcely veiled hostility. Diplomats of the US and of Great Britain tried to undermine its unity by using their influence in the Middle East to try to persuade those countries to leave the organization. There have been internal splits and rivalries in OPEC and there will be more in the future. But it has survived for sixteen years and that is a fact which reveals that the big oil producers (which are relatively few in number comprising Venezuela, and parts of the Middle East and North Africa) are fully aware of what their privilege in possessing oil means, and also realize that they can only get an equitable share in the wealth accruing from oil by promoting a coordinated policy of defence.

This oil policy maintained by the democratic governments of Venezuela since 1959 has met obstinate opposition from the widest possible variety of groups, individuals and newspapers. Communists and arch-reactionaries alike have attacked it in aggressive tones. The puppets of Moscow, whose master for a time was Fidel Castro, have accused us of submitting to foreign interests. The reactionaries have used different weapons, some of them simply wanting to make things as difficult as possible for us, and others acting in an only partly concealed way as agents of the oil companies. They claimed that we were committing a crime far worse than a series of administrative errors, which amounted to damaging the very essence of the nation, by putting up the taxes on the industry, by founding a state company, the CVP, by making a defensive alliance with the Arabs in OPEC, and finally by refusing to put up on offer, under the colonial-style system of concessions, the oil zones which remained in the country's possession. These frightened Cassandras cried out that the companies would pack their bags and leave Venezuela and turn their attention to the Middle East and North African countries, which offered tranquillity, inexhaustible deposits of oil and an unconditional welcome to foreign capital.

This artificial argument collapsed with a crash when the Arab-Israeli War broke out on 5 June 1967. This showed how 'insecure' oil production was in the Middle East and North Africa and reminded the world that Venezuela remains the basic supplier of the West's oil, and also the oil country which can guarantee most stability. A quick summary of what happened before, during and after the 'lightning war', and a rapid analysis of the rôle of oil in the unresolved Middle East crisis will bear out these statements.

Oil, hero or villain of the Arab-Israeli war?

On 27 May 1967 Egyptian troops sealed off the straits of Tiran and stopped Israeli ships from entering the Gulf of Aqaba en route to the port of Eilat. This was the port through which Israel got the supplies of oil needed for her industries, her mechanized agriculture, etc. The Gulf of Aqaba was the only way out for Israel's Red Sea trade. The Israeli government announced officially that this advance by Nasser's troops provided ample justification for a declaration of war. There was a short burst of diplomatic negotiations which raised hopes that war might be prevented by a friendly agreement worked out by neutral mediators. But these peace negotiations failed and on 5 June Israel pushed troops into Egypt, Syria and Jordan. What happened afterwards is too well known to need retelling here.

True to form, oil made its presence strongly felt in the cataclysm. It had been an underlying factor in previous Middle East conflicts, and at an earlier stage of this one, but now it came right out on the surface. The Arabs were able to use their huge potential reserves of this prized raw material as an effective lever against the USA and Great Britain, the leading allies of the Israelis.

The good sense behind this strategy can be seen most clearly if we look at the vital rôle played by oil from Middle East and North Africa in keeping the powerful manufacturing and war industries of Western Europe going. That is what I shall now do.

Western Europe consumes 8·5 million barrels of oil a day. Half that amount comes from the Middle Eastern and North African countries. Great Britain is particularly dependent on these countries, which supply 66 per cent of the oil required by its industries, its merchant navy, and the Royal Navy (figures for 1967).

Given this situation, the Arab rulers and those of the North African countries felt that the best way to take revenge on the USA and Britain, the countries which they regarded as the most active allies of the Israelis during the war, was to keep in force their boycott of oil supplies to those countries.

This menacing threat was echoed by the governments of two Arab countries in Africa. The position of Algeria and Libya was very significant because both were linked by religion as well as blood with their Middle East brothers in a 'holy war' against their 'triple aggressors': Israel, the United States and Britain. As for their oil, Algeria now produces 33·8 million tons a year, and Libya's rapidly growing production amounts to 72·3 million tons. Together they contribute 102·5 million tons of the oil consumed by Western Europe every year.

The oil companies which supply the West European market had another problem to face, namely the explosive internal situation in Nigeria which supplies Europe with 20·7 million tons of oil a year. This country was in the throes of a bloody civil war, provoked by age-old tribal feeling and the uncontainable rivalries of the country's military leaders. The Eastern region of the country, where most of the oil wells, refineries and oil shipping installations are, declared its independence of the federal government in Lagos and took the name of the Republic of Biafra.

This autonomous government demands the right to collect royalties and taxes on oil produced on its territory. So, even before the civil war between the Northern and Eastern regions broke out, the oil companies and the governments which protect them had to make the very difficult decision of whether to pay taxes and royalties to the federal government or to the new Republic of Biafra.

Beginning in the first week of June 1967, these problematic situations gave deep concern to governments and governed alike in the oil-oriented industrial nations of Europe. Britain, if it started fuel rationing, had reserves for 5½ months; in the other European countries the situation was similar if not worse. The United States found that the supply of 100,000 barrels a day from Saudi Arabia for its navy off Vietnam was cut off. For a few days the US could only supply the urgent need for oil in the Far East, above all in Vietnam, by taking it on a roundabout route by sea.

The West assumed that the oil countries would not be able to make their boycott against the Anglo-American oil companies last, and Egypt would not be able to keep the Suez Canal closed, despite the threats expressed not with reasoned arguments but in outbursts full of hate by the Arabs. In fact oil production has begun again, but at the time of writing (December 1967) the Suez Canal is still closed. So Egypt is not collecting the 1,000 million bolívares it received every year in tolls paid by the ships which use the canal. Iraq gets half its revenue from oil, Saudi Arabia 88 per cent and Kuwait 92 per cent; so they depend entirely on a constant flow of oil. In 1966 these countries were paid 11,500 million bolívares by the oil companies in taxes and royalties. Without these payments the desert would again take over completely and only sheep would be able to feed in these inhospitable regions which are desperately poor in all natural resources save oil. The North African countries depend to a similar extent on oil, which accounts for 95 per cent of Libya's exports, though Algeria does export other commodities. In Nigeria the bloody civil war means that there are bleak hopes even for maintaining oil production at its present level. All this explains why the oil companies resumed activities on 15 June 1967 in Egypt, in Kuwait (which produces 120 million tons of oil a year) and in Saudi Arabia (also 120 million tons), from which ARAMCO has already resumed supplies to the US army in Vietnam.

On 18 June 1967 the tension between the oil companies and the governments of the Middle Eastern and North African oil countries came into the open again, when the foreign ministers of the Arab League met in Kuwait. All the Arab countries were represented, including even the Sudan, which hadn't exactly been active during the war. Egypt, Syria, Algeria and the Sudan made strong demands for the suspension of all oil shipments to the West. They argued that if they limited the boycott to supplies of oil to the US, Britain and Israel, it would be ineffective because they could not prevent oil being re-exported. Saudi Arabia was harshly criticized for letting ARAMCO begin operating again. There were divergences of opinions over whether or not to maintain or to extend the boycott, among other things, but there was a general agreement to take reprisals against Britain and the USA, using their oil companies as scapegoats.

Future prospects, therefore, for the great international oil companies in the Middle East and in North Africa are not at all stable or reassuring. In the West there is speculation about the authenticity of an Israeli recording of a telephone conversation between Nasser and King Hussein, in which they agreed to announce simultaneously in Amman and in Cairo that US and British air force planes took part in the bombing of Egypt and Jordan. But the Arab peoples firmly believe that this is true. On 26 May 1967 the Arab Workers' International declared: 'When battle is joined all the oil wells, pipelines and other oil installations must be destroyed.' This extreme threat was not carried out but the position of the oil companies became very difficult during the Arab-Israeli war and the future is very uncertain for them.

The balanced and well-informed London newspaper *The Times*, in an editorial on 9 June 1967, gave a summary of the alarms which the Middle East war crisis had provided for the directors of the oil trusts and the governments of Britain and of the USA, the countries where they have their head offices. In Syria outbreaks of popular violence forced the companies to evacuate many of their employees. In Saudi Arabia the oil terminal and refinery at Ras Tamura was closed at the demand of the workers. In Kuwait the most important step taken by the government in the days leading up to the outbreak of war was to send troops to protect the installations of the Kuwait Oil Company from mob violence. In Saudi Arabia again the ARAMCO installations and refineries were ransacked by angry crowds. Iraq had to stop all production not only as a reprisal against the West but also because it knew that the Syrians would cut the pipelines which led through to the Mediterranean as they had done during the 1956 Suez crisis. Lebanon also closed the pipelines inside its borders. On 26 June, several weeks after the end of the war, Arab nationalists blew up in the port of Aden two gigantic Standard Oil tankers which were brimful of oil.

This London newspaper also made some forecasts about the difficult situation which awaited the oil companies in Arab countries after Israel's victory in the war. Quite apart from the war, and the even greater danger of popular violence against them, the companies face many problems in negotiating with Arab governments, both present ones and future ones, because it looks as if those of Iraq, Kuwait, Libya, and possibly Saudi Arabia, will soon be overthrown. *The Times* predicts that in these circumstances, 'the classic political solution will be to show Arab virtue and throw the oil companies out'.

One of the first effects of the war, according to *The Times,* will be to increase pressure from the Arab governments for better terms in their deals with the oil companies. They may possibly go as far as they did during the Suez crisis in 1956, when Nasser expropriated all the oil companies operating in Egypt. After doing that, Nasser went on to sign service contracts with Philips and with Pan American Oil. It is quite possible that Iraq may put into practice its so-called Law 80, which expropriates half the concession enjoyed by the Iraq Petroleum Company. *The Times* reaches this stark conclusion: 'There can be no doubt that the American and British companies will be at a serious disadvantage when they come to seek new concessions in the Arab world.' In view of this, countries with strong economies and technical know-how, like Italy, West Germany, Japan, France and even Spain, are willing to invest expertise and money in the oilfields of the Middle East and North Africa, and are on the look-out for future openings. This helps to explain the French government's policy of strict non-intervention during the recent war, even though French public opinion was strongly in favour of the Israelis, and even though General de Gaulle called Israel 'our friend and ally' in an official speech. Before the end of 1967 there was one concrete explanation for de Gaulle's Arab sympathies. On 23 November, after some highly secretive negotiations, France signed an oil agreement with Iraq. When the news came out, Iraq repeatedly insisted that they had signed a service contract, and that it was no concession with overtones of colonialism. So thousands of miles away from Venezuela the words and ideas of our nationalists were reproduced by the Iraqis. It was the state-controlled group ELF which signed this agreement for the French with the Iraqi government. It is worth running through the clauses of the contract to prove that it is perfectly possible to achieve through service contracts what our CVP wants to do. The contract stipulates that ELF will look for oil in an area assigned to the Iraqi state oil company where oil has yet to be found and that it will do so in the name of the Iraqi company. The explorations will last six years until 1973, when ELF will hand the land back to Iraq. If oil is found ELF can exploit it for 20 years, giving half the oil produced to Iraq. Within five years of the wells being brought into production, the administration will be 'wholly' in Iraqi hands, with the French acting only as advisers. ELF will pay 75 million francs for the initial exploration rights. This agreement shook the international oil world. Shell, BP, and Esso-Mobil, which make up the Iraq Petroleum Company together with ELF, accused their French partners of double-dealing, for not leaving them a share in the agreement. The lesson for Venezuela is that service contracts, even with particularly favourable conditions for the producing countries, are not only accepted but also competed for by the great international companies.

There is another factor which brings instability to

the oil-producing countries of the Middle East and North Africa and increases the risks for the companies from the West; the active manoeuvres of the Soviet Union in the area. The Tsars thought of the Middle East as a 'natural' zone for Russian expansion. This truly imperialistic idea has been automatically taken over by Josef Stalin and his successors. The Arab Middle East, an area which is always liable to political instability, is easy meat for the destabilization experts of the Kremlin and Cominform.

In the Arab Middle East the Russians are playing several games simultaneously. They are trying to work up further hatred against the American and British companies, which have already been discredited by their own actions. At the same time, to the detriment of the economic interests of their Arab 'allies', they are carrying out an aggressive commercial campaign to invade the Western European market for oil and gas. However it won't be a total invasion because, even though Russia is the second biggest oil producer in the world, it consumes enormous quantities itself and also has to supply the requirements of the so-called democratic republics of Eastern Europe.

On 16 June 1967 another London newspaper, the *Financial Times*, published a significant piece of information: 'The Soviet Union is promoting a strong campaign to sell oil in Western Europe, taking advantage of the Arab oil embargo and the closure of the Suez Canal.' The newspaper explained that the Russian government made offers of oil to importers in several European countries, including Britain. Britain has a long-standing prohibition on all oil imports from Russia, but now the Board of Trade was showered with demands from big industrial firms for an end to this measure. According to the *Financial Times*, the Wilson government was keeping 'an open mind' on this subject. If they maintain commercial links with Cuba in spite of the Organization of American States boycott on trade with that country, why shouldn't the British buy oil from Russia? After all, they would argue, business is business. Russia is also negotiating an agreement by which it will be able to sell enormous quantities of gas to Italy (a figure of 12,000 million cubic metres a year has been mentioned in the press) after building a gas pipeline from the Urals to the Adriatic Sea, at Trieste, with outlets in Austria and France as well. During the same period immediately after the Six Day War Russian oil was also offered to Switzerland and other European countries. Reputable newspapers even announced on 15 June that Russia had reached agreement with the Spanish government's oil trade agency, OGRO, to supply some of Spain's oil needs. It was the semi-official Spanish news agency CIFRA which first announced this. But it was soon denied by the official Soviet newspaper *Izvestia* on 16 June 1967.

In spite of this denial by official Soviet resources (after fifty years in power the régime still keeps very tight control over the whole information system), the fact is that Russian oil is reaching Spain. In confirmation of what CIFRA said, 500,000 tons were shipped from the Black Sea to Spain's Mediterranean ports. On 23 June 1967 the Soviet tanker *Leonardo da Vinci* entered port as Cartagena with 50,000 tons of oil for the refinery at Escombreras. A week earlier two other Russian tankers had landed 90,000 tons in Spanish ports. La Pasionaria and the old guard of the Spanish Communist Party still rant and rave against Franco and his régime in front of the microphones of the powerful radio transmitters on Czech or Russian soil, and the Spanish government's firing squads have just left the body of Grimau, the leader of the pro-Soviet underground political movement in Spain, riddled with bullets. But at the same time Moscow and Madrid have put out feelers for a renewal of diplomatic relations, and the Spanish state refineries are being supplied with Russian oil. What is clear is that the Soviet government is taking the opportunity of the interruption of the flow of Arab and North African oil to get a foothold in the European market so as to benefit its own selfish political and economic interests, and not 'so as to cut the ground away from under the feet of the international oil companies' as the *Le Monde* writer Alain Muncier put it in a charitable article on 17 June.

But, perhaps because it was caught in the open, the Soviet government cut short its double game, which amounted to making declarations of love to the Arabs and at the same time sabotaging their oil boycott. On 21 June 1967, the world press carried the news that the offers made by Russian commercial agents to Britain, Belgium, Norway and Switzerland had been withdrawn for the moment. But this is just a brief halt, designed to save face in front of the Arabs, in the long Russian campaign to sell more oil to the West. Russian oil sales abroad rose by 13·7 per cent in 1966 and totalled 50 millions tons; 70 per cent of that enormous amount went to Western countries.

What Venezuela has gained from the upheavals in the Arab world

Venezuela hit the front-page headlines in connexion with the Arab-Israeli conflict, as the crisis built up, during the Six Day War, and in the subsequent disputes. This was simply because of our oil potential. Because Western markets were cut off for a time from their supplies of Middle Eastern and North African oil, the governments of the West and their military high commands came to turn their attention and once again to place their faith in our known reserves of over 16,200 million barrels of oil. In fact we have much more than that. But up to now the companies have used the geophysical system, which does not reveal as much of the true extent of the reserves as the drilling system, which reaches the deepest layer of the subsoil. We must now get this method into operation and also explore the huge chunks of hundreds of thousands of hectares of unused land inside the bounds of existing concessions.

The rate of growth of Venezuelan production was forecast at 6 per cent for 1967. This is a very reasonable rate of growth to maintain; but if we were to begin to put up production by leaps and bounds again as we did from 1956 onwards, after the Suez crisis, we would simply be repeating a very costly mistake. It is quite possible to avoid it this time, because, since 1959, oil policy for Venezuela is not made only in the head offices of the companies in New York and London. Venezuela can and certainly does also play a very important part in deciding policy.

Venezuelan production went up before, during and

after the war. The fluctuations can be seen in Table 2.

Table 2 *Venezuela: increase of oil production as a result of the Arab crisis*

1967		*Barrels per day*
a) 27 May–5 June		3,364,844
b) 6 June–15 June		3,529,167
Fluctuation:	In barrels	164,323
	Percentage	4·9%
1 June–15 June		
a) 1967		3,462,971
b) 1966		3,293,427
Fluctuation:	In barrels	169,544
	Percentage	5·1%

By December 1967 Venezuelan production was as high as 3,800,000 barrels a day.

The US Defence Secretary officially ratified what the newspaper had already been saying when he said that the crisis of oil supplies would be overcome by increasing production in the United States, Canada, and Venezuela. In one statement he destroyed the unfounded cliché about the 'Caribbean reserves' by making a specific mention of the only country in the Caribbean and in all of Latin America which really has large proven and assessed oil reserves: Venezuela. In considering the possibility of increases in production in the US and Canada, we must bear in mind three facts: 1 the very high level of domestic consumption in both these countries; 2 the high production costs, which are double those of Venezuela, and three times as much as those of the Middle East; 3 the conservationist policy maintained by Canada and the US, which forces them to be very sparing about increases of production. Mexico was not mentioned at all, for it now consumes most of its own oil production and has very little left over to export. What should we conclude? All this provides a new argument which Venezuela must use as much as possible to press home its long campaign to put an end to the quotas and import restrictions imposed on our oil by the US. Venezuela is the only country which can assure a steady and secure supply of liquid fuels for the USA and Western Europe both in peace and in war. The importance of the Middle East and North Africa as great exporters of oil has also been brought out by the war. So our development of close links with these countries by the creation of OPEC has been proved right once again as part of our overall strategy to maintain the price of our basic export commodity. The example of OPEC is now being followed in other spheres; for example, in the first week of June 1967, the representatives of four great copper-producing countries, Chile, Peru, Zambia and the Congo (Zaire), met in Lusaka. These countries, put together, produce about 2,050,000 tons a year, which is three-quarters of the copper consumed by the Western world. They want to set up an international organization with the same aims as OPEC's, namely to stabilize the prices and control the production of their commodity.

It has also become obvious as a result of the swift Arab-Israeli war that the idea that the oil companies might leave Venezuela if the government did not stop its nationalist policies in defence of our oil is a complete myth invented by those who wish to frighten the nation. They won't leave, not only because of the value of their investments in Venezuela, but also because they've got a very hot potato in their hands in the Middle East, which might burn their hands and could yet burn up their oil wells and refineries. Venezuela is the only big oil-producing country and the top exporter which enjoys real political stability and freedom from social conflict, and thus can guarantee a continuous supply of oil.

I come finally to what is perhaps the most important lesson to be learnt from the Arab-Israeli War. The far-sighted and patriotically motivated policies of the democratic governments since 1959 in refusing to offer new concessions on the open market and in setting up the CVP and supporting it actively have been proved absolutely right in the light of the conflicts which face the international companies in the Middle East and North Africa. There are now many Venezuelans who realize the strategic importance of our oil as a result of the information which has been made available to ordinary people in the press and on television. If we had put our oil resources on offer at a knock-down price through the concession system, we would have killed the very promising gosling hatched by the goose that laid the golden egg, or at the very least have left it hardly breathing, to borrow Senator Uslar Pietri's simile. There is good evidence now to back up our argument that, with the formula of service contracts, Venezuela can get more benefit from our oil than if we had already signed away most of our reserves in concessions. We aim in the near future both to get more in taxes and, more important, to get a significant share in decisions on the expansion and policies of the companies in general, and above all in relation to the new areas they will be assigned for exploration and exploitation. The CVP has already carried out one vital part of its basic function, by being the first Venezuelan company to take part in the process of exploring for, exploiting and marketing some of our oil and its by-products. It has still to begin its most important task, namely to represent the country and the government in the brand new organization which will be set up to deal with the administration of the service contracts.

To sum up, three basic points can be made about the effects of the lamentable armed hostilities between the Arabs and the Israelis, which have cost so much in terms of human lives and human suffering, on Venezuela and her economic development:

1 Venezuelan oil has been proved to be the key factor needed to keep industrial production going and to guarantee military security in the West.
2 Venezuela has come out with a strengthened position for its links with the international companies and its commercial relations with the USA.
3 The facts in all their obstinacy, as the British would say, that Venezuelan oil policy since 1959 has not been capricious or inconsistent; it has been worked out realistically and with a firm basis in unshakeable principles of patriotism and nationalism.

Oil: an Indispensable Source of Energy

Oil, Ruler of the World

My heading comes from a short French book. The French haven't produced many illuminating or well-informed books about the function of oil in modern life. The Anglo-Saxon countries' literature on this subject does really bring out the significance of this all-important source of energy for peace-time industrial production and war machines alike. But the phrase coined in the French book is a good one: 'Oil, Ruler of the World'. As industry has become increasingly more and more mechanized, oil has turned out to be the cheapest and most efficient source of energy for motors in factories and elsewhere, for aeroplanes, cars and all other forms of transport.

There has been a progressive change-over, at times slow and sometimes with spectacular acceleration, from coal to oil as the fuel used to run the industrial and military complexes of the highly-developed countries. During the First World War the main fuel which powered the opposing armies was still coal. The development of aviation, of tanks for the armies on land, and of diesel-powered ships marked the beginning of the period in which oil has come to rule the world; that period still lasts today and will be with us in the foreseeable future.

There was another basic change in addition to these ones. Heavy industry discovered that coke as a fuel was obsolete and changed to liquid fuels of many different kinds derived from oil. Moreover, with the development of the petrochemicals industry, oil became the basic raw material for a vast range of products.

A peculiar feature of the oil industry is that the big producers are very few in number. Apart from the countries which are also subcontinents, the United States and the Soviet Union, the areas which produce oil in significant amounts are simply the Middle East, North Africa, and Venezuela. The rest of the world produces little oil or none at all.

Cut-throat competition developed between the big companies over the possession of oil-bearing tracts of land as a result of this peculiar situation, and because it was found that high profits could be made from oil-wells. The Latin American Indians who called the black oil which oozed out of the subsoil in some countries 'the devil's excrement' had strange foresight. The history of oil is a catalogue of the problems it has caused for the small countries which produce it. It has been responsible for revolutions, coups, international wars, traffic in political influence, and bribery in the administrations of the countries which are lucky enough to possess large deposits.

The oil business

It is widely known that the business of the oil companies is the most profitable of all investments in the capitalist world. The companies which operate in the Middle East, in North Africa and in Venezuela have reaped profits equal to several times the total value of their investments. This is not made fully clear in the figures they produce when they distribute dividends to their shareholders, though the high rates they pay to the coupon-clippers gives an indication of the size of their profits.

Fiddling the balance sheet can be done in many different ways in the most highly-integrated industry of them all, which embraces everything from the oil-wells to the petrol stations which distribute the finished product, including refineries, pipelines and shipping lines. The international companies are like gamblers who make a killing on many tables at the same time. There have, however, been a few occasions on which those who have made fortunes from the oil business have candidly admitted the attractions of the profits they creamed off from the oil companies' dividends.

One example is the autobiography of Nubar Gulbenkian, the son of the famous Armenian who made the most valuable collection of concessions in the history of oil. Nubar Gulbenkian writes:

> Before I weep over the tragic fate of great capitalists (and I make no exception for my own family in this), I shall examine some figures. Between 1914 and 1953 the Gulbenkians consistently had between £500,000 and £1 million invested in the Middle East. In the years since 1955 alone these investments have produced £5–6 million a year. The profits of the oil groups have obviously gone up in the same proportion as those of the Gulbenkians' investments

at 5 per cent. Everyone who has business interests in the Middle East and looks calmly into the future must realize that if his investments are bound to dwindle gradually, even now capital invested in that part of the world is generally amortised in one or two years. Until the very end it will be a profitable place to invest. (*Nous les Gulbenkian. Les aventures dorées du petrole*, Stock, Paris, 1965, p. 219.)

The rebellion of the little countries which produce oil

The three years from 1945 to 1948 have a supreme importance in the history of the oil industry from the point of view of the developing countries which produce oil. During this period Venezuela was governed by Acción Democrática and I and later Rómulo Gallegos were the national Presidents. Venezuela was then the world's leading oil exporter, and the second biggest producer after the USA. Before we came to power we had already worked out our tactics, and a strategy which was revolutionary and at the same time realistic, for dealing with the double problem which faced the country: first, the existence of both government and society depended almost entirely on the foreign exchange earned by oil exports; and secondly, the nation's share in the enormous wealth of its subsoil was quite scandalously small in comparison with the profits made by the concession-holding companies.

We never proposed a decree or a law to nationalize the industry, and we decided to recognize the validity of the concessions granted by previous governments. We fully realized that the country was not equipped to take an active role in three sections of an industry which by then was already producing over 1 million barrels a day: the refining, transport and storage of crude oil and its by-products. What was most important was not to rely on the word which some people simply turned into a magic formula: nationalization.

What the Venezuelan government did in 1945–8 amounted to the following:

1 It stopped granting new concessions and thus put an end to the desperate scramble for the country's subsoil which had been going on.
2 For the first time in a country in which the concessions system held sway the government took some of the royalties it was due in the form of oil, and put it up for sale on the free market.
3 It raised the country's share in oil profits to 50 per cent. This was the famous fifty-fifty formula which has spread like wildfire through the Persian Gulf countries and through all the developing countries which produce oil in significant quantities.

This move towards a nationalistic policy was reversed in Venezuela during the dictatorship which began on 24 November 1948 and was overthrown by the nation as a whole on 23 January 1958. During that murky era new concessions were granted and the state's inspection of the profits of the companies was stopped. The closure of the Suez Canal in 1956 brought a rapid jump in Venezuelan oil production and an orgy of public spending by the irresponsible and unscrupulous group which had taken power by force.

The Third World's first success

In 1959 Acción Democrática returned to power in Venezuela and I became President again.

The economic and fiscal prospects were bleak. The country had been burdened with a public debt of thousands of millions of bolívares. Unemployment was running at a high rate. The prosperous years when Venezuelan oil was much in demand in the developed world because of the closure of the Suez Canal, had come to an end.

In 1960 the oilfields of Libya burst on to the scene with a bang. Under the very best possible conditions for the companies that country ruled by an inefficient and corrupt king began to produce thousands and later millions of barrels a day of cheap, high quality

Table 3 *Average Daily Production (in barrels)*

Country	*1969*	*1970**
United States	9,210,200	9,525,000
Soviet Union	6,580,000	not available
Venezuela	3,591,000	3,699,000
Iran	3,374,700	3,880,900
Libya	3,110,700	3,763,900
Saudi Arabia	2,992,700	3,330,000
Kuwait	2,575,500	2,669,100
Iraq	1,526,300	1,550,000
Canada	1,126,320	1,224,749
Algeria	943,000	976,500
Indonesia	764,400	947,800
Mexico	405,100	410,110
Qatar	355,400	363,900

*First 6 months only (Source: *Petroleum Encyclopaedia*, 1971)

(low sulphur content) oil for the European markets. The world was literally flooded with low-price fuels.

With the scarcely concealed intention of leaving the producing countries to suffer the consequences of this situation, the international companies made a unilateral decision to lower the reference prices on which taxes are calculated in the Persian Gulf and the Gulf of Mexico; this substantially reduced the revenue collected by the Middle East countries and by Venezuela. In 1959 the price at which Venezuelan oil was quoted was pushed down by amounts which varied from 5 US cents a barrel. In the Middle East the maximum price cut was 18 cents a barrel.

The oil-producing countries realized that it was the moment to stop complaining about the way the oil companies treated them almost as if they were their colonies. The idea of making an agreement to take up a united stand for equitable distribution of the profits made from their oil began to gather momentum. From 10 to 14 September 1960 representatives of the governments of Iran, Iraq, Kuwait, Saudi Arabia and Venezuela met in Baghdad. This historic meeting laid the foundations of the Organization of Petroleum Exporting Countries (OPEC), which has become world-famous. For the first time the countries of the despised Third World were fighting the monopolistic tendencies of the international companies and the industrial states which protected them with a sellers' cartel of their own.

From then on there was a bitter struggle to stop the buyers and the concession-holding companies being the only people who could control the price of oil. The producing countries were also in a position to play a decisive part in this, as they now began to do. A Venezuelan, Dr J. P. Pérez Alfonzo (who had been my Minister of Mines and Hydrocarbons), and an Arab, Sheikh Tariki, led this ambitious, yet realistic crusade.

From 1959 on Venezuela was not only the moving spirit behind OPEC. We also brought back the policy of no more concessions and set up our own state oil company, the Corporación Venezolana del Petróleo (CVP), to which millions of hectares were assigned by the government. This coherently nationalistic policy was bitterly opposed by the multinational companies in the country. They did this, however, in an underhand way and refused to take the blame for harassing the government.

They applied overt pressure to some extent through the Cámara del Petróleo, which was affiliated to the powerful FEDECAMARAS (Federación de Cámaras y Asociaciones de Comercio y Producción). The leading members of the Cámara del Petróleo are Venezuelans drawn from the ranks of the top managers of the oil companies which operate in the country. Their criticism is quite balanced, not bitter. In the companies' overall plan, the task of making strong provocative attacks was left to journalists in the press, on the radio and on television.

The aim of this well orchestrated publicity campaign was to build up a feeling of insecurity in the country by putting the blame for two things on the government's not just mistaken but criminal policies. These were, first of all, the companies' reluctance to make new investments (this ignorant and insensitive campaign went so far as to suggest that the foreign companies might leave the country altogether). Secondly they argued that, by promoting OPEC and forming a defensive alliance with the oil countries of the Persian Gulf and North Africa, the government was encouraging the rapid increase of production in areas outside Venezuela which might well deprive Venezuela of present or future markets for her oil. They printed a picture of Venezuela as a sacrificial lamb destined, because of her government's stupidity, to be roasted and eaten by the kings and emirs of the Middle East.

In 1970 the text of the letter of resignation submitted by the Creole Petroleum Company's public relations department's sub-manager, Dr Francisco Alvarez Chacin, on 31 August 1960 was published. The information it contains gives an insight into the methods used by the companies to manipulate public opinion through their public relations departments and through the Cámara del Petróleo: '. . . the disagreements have become more serious as a result of the latest price cut dictated by ESSO Export [another subsidiary of the company which owns Creole] and the defensive measures taken by the Venezuelan government, above all Resolution 994 of the Ministry of Mines and Hydrocarbons.'

He goes on: 'I particularly object to a motion passed by a majority of the executive committee of the Cámara de la Industria del Petróleo which sets in motion a publicity campaign demanding among other things the withdrawal of this government resolution, just in time to weaken the position of the Venezuelan representative at the forthcoming Oil Congress in Beirut.'

Alvarez Chacin admits that, after his verbal threat

to resign from his job, the Creole Petroleum Corporation promised to allow him to stop the campaign to discredit the government's policies which had already been put into action by the Cámara de la Industria del Petróleo. But Alvarez Chacin did not go back on his decision (F. Alvarez Chacin, *Secretos petroleros contra Venezuela*, Caracas, 1970, 17–18).

Some individuals were also among the resolute opponents of the oil policies. One of the most obstinate of them was the talented and widely known author, Dr Uslar Pietri. I do not wish to suggest that he was acting in concert with the oil companies. There is no doubt that it was lack of information on the solid long-term objectives or our oil policies which led him to make violent speeches against them.

In 1966, in a speech in the Senate on 18 May, he was still repeating the old myth that OPEC was damaging Venezuela's interests and had simply become a useful lever for the Middle East countries. It was rather sad to hear him say in that speech: 'What I am wondering is whether our membership of OPEC and our oil policies in general are meant to assist the development of Iran or to help Venezuela.' In his book *Petróleo de Vida o Muerte* (Caracas, August 1966) Ulsar Pietri reprinted much of what he said and wrote against a policy which no tolerably well-informed Venezuelan thinks of questioning today.

A double game

The oil cartel was playing a double game. In Caracas it hinted at crooked dealings in OPEC by Middle Eastern or North African governments. In Tehran, and the other Arab capitals, it developed intrigues against Venezuela, by suggesting that our country and our government were using OPEC as a cover for our attempts to displace Arab oil in the world market.

Rouhani, the secretary of OPEC and adviser to the Shah of Persia, brought to light these dirty tricks by the oil companies in Caracas in May 1963. In a very blunt statement he accused Howard Page, Standard Oil's top representative in the Middle East, of having said in Persia that the Venezuelan subsidiaries of his famous trust had enormous reserves they would not exhaust even if they got the duration of their concessions doubled. As Rouhani said without ever, so far as I know, being contradicted, Page had gone on to say that Venezuela only wanted to make use of OPEC to raise fuel prices so as to take back markets that had been won from them by Persia and the other Arab countries.

In order to carry out its policy of strengthening ties with the Arab countries, the Venezuelan government decided to set up embassies in that area. But this highly logical policy from our point of view disturbed the Foreign Ministry of Israel, a country with which we also maintained very friendly diplomatic relations. Golda Meir, then the Foreign Minister of that country, expressed Tel Aviv's deep concern when she visited Venezuela.

When we met in my office, Mrs Meir said, in her aggressive tone of voice, which possibly dates from the time when she was active in US labour unions, that Israel considered our move towards the Arabs an unfriendly act. My reply was firm but restrained. The rulers of Venezuela are Venezuelans, not Israelis, she was told, and the defence of the national interest is what decides the government's foreign policy.

It was an unhappy occasion. There was another difficult moment in similar circumstances, this time caused by the Arab delegates at the public opening ceremony of the 21st OPEC meeting, held in Caracas in December 1970. The Arabs threatened to walk out if the Israeli ambassador remained in the hall, and deadlock was only broken by the Israeli diplomat's quiet departure. It is very difficult for governments such as our own to maintain a position of neutrality in the bitter struggle between Jews and Muslims.

A resounding victory

During 1970 the situation of the world oil market began to change beyond all recognition. There was a massive increase in consumption. As a result of the Six Day War the Suez Canal was closed to tankers indefinitely. In Libya a headstrong nationalist régime had taken over from the obliging monarchy of King Idris.

The OPEC meeting in Caracas in December 1970 and the legislation putting up oil taxes, which was approved by Venezuelan Congress in the same month, were the prelude to the famous meeting at Tehran in early 1971. For the first time the arrogant international oil companies had to give up their conveniently divisive system of carrying out separate negotiations with each country in turn; they had, very reluctantly, to admit that they had finally reached the point at which they had to negotiate with OPEC as a whole, at least in relation to the Persian Gulf region.

The talks were slow and laborious, with a few dramatic moments. At one moment the Shah of

Persia reminded the companies that Venezuela had already shown the way forward and said that the Middle East oil countries would follow that lead if they could not reach a bilateral agreement. In other words the different Arab countries would get legislation passed to put up the taxes on oil and to ensure that they were in a position to fix the reference prices for tax purposes themselves.

In the end the international companies gave way. The agreements signed in Tehran put taxation up from 50 per cent to 55 per cent, which was still 5 per cent less than the rate approved by Venezuelan Congress in December 1970. The reference prices were put up by 35 US cents a barrel, and are to go up a further 11 per cent each year until 1975.

A difficult agreement

Soon after this came Libya's confrontation with the international oil companies. The conflict was even more involved and harder fought than the negotiations at Tehran. The extremely nationalistic Colonel Gaddafi and his advisers gave full rein to their hatred and bitterness against the companies which had mercilessly exploited their country under the corrupt régime of King Idris. It was freely remarked that moderating influence was brought to bear on Gaddafi by the President of Egypt, Anwar Sadat, who was concerned about the economic aid his government gets from Libya.

Gaddafi and his advisers at one stage seemed ready to take over the oil-wells and keep them closed indefinitely rather than settle on new terms for the exploitation of the subsoil. But in April 1971 these difficult negotiations came to an end with an agreement which gave Libya a 90-cent rise in the reference price per barrel, and further rises each year until 1975; taxes were fixed at 55 per cent and Libya got additional payments for retrospective claims.

There was also a confrontation in 1970 between France and her former colony Algeria, now her main supplier of liquid fuels. President Georges Pompidou's government finally accepted Algeria's demands for a 51 per cent share, and thus control over decisions, in the French oil companies, most of which are state-owned.

One of the main factors which contributed to the all-round success of the producing countries' demands was the constantly accelerating growth of fuel consumption in the Western world.

European consumption went up from only 1·2 million barrels a day in 1950 to 12 million in 1970, which is only just under what the USA consumes. It is forecast that by 1980 Europe will have overtaken the US and will need 23 million barrels a day. Japan's consumption of oil has risen spectacularly, keeping pace with the huge strides made by industrial production there. In 1950 it needed only 100,000 barrels a day, but in 1970 it consumed 3·7 million, which could be 10 million by 1980. Furthermore, 'The prospects for economic and social progress in the developing countries depend above all on an increased consumption of oil.' (*Oil Power Affairs*, July 1971.)

Rising consumption

A few facts and figures suffice to indicate how the demand for oil in the industrialized countries will snowball. Oil consumption in the non-Communist countries has jumped from 10 million barrels a day in 1950 to 59 million in 1970, and could reach 67 million in 1980. In the United States oil consumption rose from 7 million barrels in 1950 to 15 million in 1970, with a 21 million forecast for 1980.

These increases in consumption have, naturally, been met by rapid increases in oil production. Production in the Western hemisphere has more than doubled, going from 8 million barrels a day in 1950 to 18 million in 1970; the corresponding increase in the Eastern hemisphere was from 2·1 million barrels to 21 million.

On the present calculations the reserves of the United States will only last for about 12 years. However the combined reserves of the Middle East and North Africa will last between 60 and 70 years. In 1970 the oil produced by all the OPEC countries amounted to 22 million barrels a day, which accounted for 90 per cent of the non-Communist world's oil trade.

Despite the discovery of new oil deposits in Alaska, the North Sea, the Far East and elsewhere, it is obvious that the Western hemisphere will continue to depend heavily on oil from the OPEC countries to satisfy its ever-increasing needs.

Without a doubt the US will have to import more oil. At present imports account for almost 11 per cent of US domestic oil consumption, and of those imports 55 per cent come from OPEC countries, above all from Venezuela. This percentage should rise considerably in the next decade, even taking into account what Alaska will be producing.

In this context the dramatic confrontation which

took place in Libya in the summer of 1969 between the oil companies and the producing countries takes on its true significance.

Libyan production grew from nothing in 1960 to 3·7 million barrels a day at the time of this confrontation, which put it not far behind the production figures of the most important Persian Gulf countries, Persia and Saudi Arabia. Libyan oil, which as I have already pointed out has a very low sulphur content indeed, is very well placed, within easy reach of the leading oil-consuming countries in Western Europe, to compete on excellent terms with Middle Eastern oil. Its competitive edge was of course enhanced by the closure of the Suez Canal after the Arab-Israeli War.

Libya and her new image

In 1970 Libya supplied nearly 30 per cent of the oil required by Western Europe. The international companies took advantage of the corrupt and incompetent monarchy to milk the Libyan deposits at a great rate. But this changed dramatically when the king was overthrown and a government which would stand up for the nation's one interest came to power in Tripoli. The conflict between the international companies and Colonel Gaddafi's government had some very tense moments. At one stage the Libyans said quite simply that they were prepared to seal off the wells and stop any oil getting out at all.

The Libyan conflict came to an end in September 1970 with a virtual surrender by the multinational companies and their numerous offshoots. The companies had to accept an increase of 30 per cent in the reference price per barrel, with successive 2 per cent rises every year until 1975, and a tax increase from 54 per cent to 58 per cent.

The little-known, sometimes even despised, group of oil-producing countries had won the first battle for themselves and for the other developing countries against the close-knit political and economic block formed by the rich countries, their governments and the oil companies.

Positive results

To show what they achieved we have only to give an estimate in millions of dollars of what the producing countries won in their struggle with the countries which exploit and burn their oil.

There are figures which reveal what the oil countries got out of their rebellion at Caracas, Tehran, and Tripoli. The revenues of the small countries which produce oil only amounted to 7,000 million dollars in 1970, but by 1975 will reach 18,500 million dollars. Without their hard bargaining, after a revolt worthy of Spartacus, their revenue would only have totalled 10,000 million dollars in 1975. The consuming countries have had to sit back and suffer a corresponding increase in the amount they pay for the oil which they import and without which their whole industrial system would come to a halt. In 1975 Europe will pay 5,500 million dollars more for its oil than in 1970, and, in the same 5-year period. Japan's payments will go up from 2,500 million dollars to 4,000 million.

Acres of print were devoted on behalf of the international companies and the governments of the industrialized countries of Europe to protests against these price rises. It was forgotten, how, for many years, but above all during the 1960s, the companies distributed enormously high dividends to their shareholders and the governments and industries of Western Europe enjoyed the benefits of a cheap supply of oil.

Everything worked out satisfactorily, because in the world of high finance there is always a tailor-made solution for every problem. The companies reduced the dividends they paid to their rich coupon-clippers and the governments let the consumers suffer the effects of the higher prices. If that contributed to fuel inflation, which was careering about Europe like a car out of control on one of its motorways, so much the worse for the consumers.

Figures can be produced to prove irrefutably, as the Economic Department of OPEC did in October 1970, that the profits made from oil were distributed among

Table 4 *Cost Factors for a Barrel of Oil in Western Europe*

	US Dollars	*%*
Cost of production	0·19	1·3
Cost of refining	0·20	1·4
Storage, distribution and selling	3·18	22·2
Freight by tanker	0·41	2·9
Net profits for the oil companies	0·32	2·2
Taxes in the consuming countries	9·18	64·0
Payments to the producing countries	0·87	6·0
Total:	14·35	100·0

(Source: *Economic Department of OPEC*)

three groups in the following order of precedence: firstly, the European governments which got the lion's share through their taxes, secondly, the companies which produce, refine and market this product, and thirdly, and a very poor third at that, the countries whose mineral wealth was being sucked away. This can be seen in the revealing table above.

An enormous increase

The people who have really lost out because of the oil price rises are the developing countries which consume but do not produce oil. The impoverished Third World will have to pay 1,000 million dollars more for its oil at the higher prices in 1975, which is a big rise over the 2,100 million dollars it had to pay in 1970. The balance of payments deficits of those countries will become far worse than it is now, and their struggle to get free of the shackles of underdevelopment will be made even more arduous.

These inevitable consequences prompted the United Nations Commission for Trade and Development (UNCTAD) to put forward a valuable and practical proposal for a fund to be set up by the oil countries in order to alleviate, at least in part, the new burdens placed on the shoulders of those destitute countries by the higher price they have to pay for their indispensable oil imports.

In late October and early November 1971 representatives of the 'Group of 77' (which now has 95 members) met in Lima, in order to prepare the Third World countries' position for the Third UNCTAD Conference, which was held in Santiago in April 1972. This group brings together representatives of the developing, or, alternatively, underdeveloped, countries of Africa, Asia and Latin America.

The group's basic aim is to force the rich, industrialized countries to pay better and more stable prices for the raw material exports of the Third World. In front of an audience of representatives of countries which were protesting against the exploitation by the developed countries, the Iranian Economics Minister, Hushang Esasi, appealed eloquently to the proletarian countries to unite in multinational groups on the OPEC model in order to achieve better prices for their raw materials from the industrialized countries which are their customers. He reminded his audience of the February 1971 Tehran agreements, which he described as the developing countries' first victory in their struggle to get a just price for their raw materials.

The Peruvian representative summed up the Iranian Minister's argument in a final motion, which was backed up enthusiastically by the Foreign Ministers of Algeria and Cuba. The motion encouraged the developing countries to reach agreements along the lines of OPEC so as to be able to hold out for a just price for their raw materials in negotiations with their customers, the industrialized countries.

With some accuracy it has often been said that history is full of irony. This occasion had, for me, many ironic echoes. The policies of my government, which had been accused on any number of occasions of selling the country down the river to imperialism, were fully upheld at the Lima Conference and adopted by the noisiest opponents of 'neocolonialism'. It should be remembered that, from 1960 onwards, Venezuela, much of the time single-handed, led a crusade to organize OPEC. The moral of the story is simple: all the industrialized countries, from East to West, including the United States, the Soviet Union, and the Common Market countries, have the same basic aim – to sell their products at high prices and buy their raw materials at low prices. Our oil policies were not inflammatory rhetoric with demagogic tendencies, as some opponents said, but amounted to a serious and realistic attempt to come to grips with these problems.

There is only one practical way, despite all the well-meaning efforts of UNCTAD, to bring about real changes in this iniquitous situation. That is by trying to unite the Third World countries in multinational groups in order to break out of the industrialized countries' cast-iron system for bringing down the prices of the raw materials they have to buy and overvaluing the capital goods they sell to the underdeveloped and developing countries.

This is an accurate diagnosis, but the remedy is difficult, though not impossible, to apply. The partial failure of the agreement of copper-producing countries, organized along the lines of OPEC, gave indications of some of the practical difficulties. Oil is the only product which is absolutely indispensable for the developed countries. The industrial machinery of the modern world simply cannot run without this source of energy, the reserves of which are distributed unevenly across the globe.

For this reason the relatively few big oil producers in the Middle East, North Africa and Latin America have been able to turn the tables completely on the developed countries. The producing countries, as the sellers, have an initial advantage, and now it is they and not the buyers who take the decisive steps to fix the price of the oil extracted from their subsoil. But this situation does not hold good for other mineral, agricultural, or partly-processed products exported by developing countries.

Oil will continue to rule the world

Has this been a Pyrrhic victory? Won't the powerful economic and political ruling groups in the West be able in the future to force the oil countries to give way again and accept lower prices for their product?

An emphatic *no* can safely be given as the answer to this question; there is no chance of the buyers being able to impose price cuts on the producing countries. It is quite plausible to argue that the technical achievement of putting men on the moon suggests that new alternative forms of energy are likely to be developed to replace oil, perhaps by harnessing the sun's rays, or by putting atomic power to the peaceful purpose of producing energy in as yet unforeseeable quantities. This was the argument a Venezuelan Senator developed recently in Congress.

But in fact this argument does not really hold water. The basic idea which the late Secretary-General of the United Nations, U Thant, developed and was able to synthesise, does not seem as yet to be applicable to oil, nor will it be applicable in the future so far as we can tell. This is how U Thant put it: 'Today the most surprising feature of the developed economies is their ability to obtain, in a very short time and in suitable quantities, any kind of resources they require. The availability of resources no longer places limits on decisions; needs in fact give birth to resources. This is the basic revolutionary change, perhaps the most revolutionary development within living memory.'

In relation to oil it is not the case that 'needs give birth to resources'. The geological and economic experts, and also, more important still, the directors of the great international companies, openly admit in public that the high cost of nuclear energy prevents it from becoming a substitute for liquid fuels. The companies are sinking bore-holes in the jungle of the Amazon and in Indonesia, investing enormous amounts on underwater explorations, and ordering gigantic supertankers, to be delivered by the ship-builders within the next ten years. Oil seems likely to go on ruling the world until the unforeseeable time when the deposits run dry.

One alternative

At the same time as the capitalist world of the West is searching frantically for new deposits of oil it is also beginning to get worried as the end dates of the concessions granted by the important producing countries draw near.

Venezuela's concessions will revert to the state in 1983–4. Those of Iran will come to an end in 1980. Saudi Arabia and Iraq seem to be ready to terminate the concessions granted by previous governments quite soon, even though the expiry dates are respectively 1999 and 2000. 'The Arabs say quite openly that the Middle East countries won't be able to wait that long,' said that balanced and well-informed British magazine, *The Economist* (31 July 1971).

The countries which own the deposits have two

alternatives when the concessions revert to them. One alternative is to take full control of all the different stages of the oil business. The other – probably more realistic – alternative is for the producing countries to use their own technical and material resources to produce the oil, and then to get favourable conditions in contracts signed with the international companies for transport and marketing. This way they will not have to divert funds away from the urgent tasks of developing their own economies and abolishing poverty to finance expensive shipping lines. They will also have to drive a hard bargain over the decisive element in international commerce, the formation of stocks, with those who have world-wide distribution networks at their disposal.

It is often asked, in connexion with the relative scarcity of oil in the Western world, 'Couldn't Russia come to the rescue and supply more fuel to the non-Communist countries?' There would certainly be no ideological difficulties from the Soviet point of view, for their greed for hard currency has no limits. But the Soviet Union is also in trouble because of the undeniable fact that its consumption of oil is rising faster than its production. A quick look at the Communist countries' oil situation will bear this out.

The Soviet Union and its oil needs

In the early 1960s the Soviet Union seemed to be on the point of launching a great campaign to sell oil and its by-products in the European market and even in Latin America. But its plan for expanding exports were checked in the light of up-to-date estimates of its known oil reserves and the ever larger increases in domestic consumption. According to figures produced by the Petroleum Press Service (January 1970), and by David Floyd in the *Daily Telegraph* (London, 7 August 1970), the Soviet Union is well on the way to becoming a net importer of oil. Consumption of oil in Russia has gone up so fast (from 98 million tons in 1957 to 533 million tons in 1970) that its reserves are in danger. It is forecast that by 1980 the Soviet Union will have to look abroad for 2·5 million barrels a day.

It is also a well-known fact that Russia has told its COMECON partners that they must look further afield for their fuel supplies, because in the future Russia and Rumania will not be able to satisfy their requirements. Moscow has in fact given the governments of Czechoslovakia, East Germany and Poland the go-ahead to negotiate oil purchases from Iran and Iraq to cover some of their requirements. But they face a difficult problem because, outside COMECON, where barter agreements are the order of the day, payment is demanded in strong currencies, like dollars or pounds sterling.

Rivalry with Mao

Western Kremlinologists have drawn two conclusions from this preoccupation of the Russians with the possibility that in the near future they may cease to be able to cope with their own fuel needs.

The first conclusion is that the bitter dispute between Moscow and Peking, which has opened a vast chasm in the once monolithic Communist bloc, must be affected by the heaven-sent, or perhaps it might be better to say hell-inspired, presence of oil. The Soviet Union has successfully opened up production in

Eastern Siberia, as less oil flows from the oilfields around Baku and in the Urals/Volga region. The potential deposits in the new area close to the Chinese frontier are enormous.

The fact that this area was seized by Tsarist Russia from the Chinese is still a very sensitive issue in Peking. Now that it is known to contain considerable oil deposits, Siberia could become a key element in the deadly rivalry between these two countries which is normally hidden behind a facade of ideological warfare between 'pure' Marxist-Leninists and 'revisionists'. Now that they have reached the age of pragmatism, the Russians are much closer to the capitalists of Japan, with money to invest in the expensive explorations in Siberia, than to their poor comrades in Peking. Japanese oil companies have already begun an ambitious search for oil in the Siberian subsoil.

The second consequence of the Russians' oil problems is connected to their active interference in Middle Eastern politics. It was not just their dislike of the Israelis which made the Soviet Union send arms to the United Arab Republic and support it during the Six Day War. Nor were they simply following in the footsteps of the Tsars who considered the Middle East a 'natural' target for Russian expansion: Moscow plots, and flatters and sometimes threatens the governments of the Middle East countries because it badly needs new sources of oil to replace its own dwindling reserves.

The Western countries, including Great Britain, the country which has been active longest, ever since its imperial heyday, in that area, now face a powerful challenge in the region where their word was law for many years. The US government's concern about Soviet offensives in pursuit of Persian Gulf oil was publicly expressed in early 1971 with considerable clarity by President Richard Nixon and his advisers Henry Kissinger and Peter Flanigan. As Flanigan put it with perfect sincerity: 'If the USSR manages, through its close links with the Arab world, to get its hands on the Middle East's oil, the US will lose much of its influence in Eastern Europe and in Japan.'

The influence of oil in Latin America

The demand for oil in Latin America is snowballing as the area's economies develop and change. The sub-continent consumed just under 2·5 million barrels a day in 1970, and by 1980 this figure should go up to 3·2 million barrels, which means a 30 per cent rise. This rapid growth of oil consumption takes account not only of development, but of the 'economic changes' which are forecast for this area in the next decade.

The figures for the consumption of oil, both as a source of energy and as a raw material for industry, will go up vertically, and, as Latin America achieves fuller economic integration, it is to be expected that countries which produce little or no oil will come to draw most of their supplies from Venezuela, the only big producer and exporter of oil in the area.

The oil production of Latin America excluding Venezuela is relatively small when compared with big producers like the USA, the USSR, the Middle East and North Africa. In 1969 the only important producer was Venezuela, with 168·1 million metric tons (1 ton = 6 barrels). The figures for the other countries were: Mexico 21·4 million metric tons, Argentina 18·1 million, Colombia 10·7 million, Brazil 8·3 million, Trinidad 8·1 million tons, and finally Peru, Bolivia, Chile and Ecuador, which, put together, produced 7 million.

Almost all the Latin American countries have state oil companies, which will coordinate the joint campaign of their member countries to get more effective control and use of their oil resources.

The present situation and future prospects

A quick look at the situation in the Latin American countries one by one will tell us the basic facts about the present situation and future prospects for oil production in this area.

Venezuela: In December 1970 the legislature passed a resolution which increased oil taxes from 52 per cent to 60 per cent and authorized the executive to fix the reference prices, on which taxes are calculated, unilaterally. After that Congress passed two more nationalistic measures: a law nationalizing gas, and a law to ensure that the oil-wells and installations return to Venezuela in 1983–4, when the concessions run out.

Under the terms of this second law the oil companies have to deposit 10 per cent of the annual cost of depreciation of their investments in the Banco Central each year, to make sure that the oil-wells and other installations will be returned to the country in good condition. The government may use these funds for social development. The sums paid by the companies cannot be deducted from their tax assessment. Led by the top four, Shell, Standard Oil, Mene Grande and Mobil, the oil companies sent a plea to the

Supreme Court for some clauses in these laws to be considered null and void, because they were deemed unconstitutional by the companies' legal departments.

This increase in taxes and reference prices produced additional revenue to the value of 440 million dollars for Venezuela in the first year it was enforced.

Here are the statistics for oil produced, refined, and exported in the two-year period, 1969–70:

Table 5 *Oil Statistics for 1969–70, in Millions of Barrels Per Day*

	1969	*1970*	*% change*
Production of crude oil	3,632	3,754	3·4
Oil refined	1,156	1,292	11·8
Domestic consumption	193	200	3·6
Crude oil exported	2,476	2,435	1·7
Refined products exported	935	1,035	10·7

Venezuelan wealth

Our country's reserves, not counting the deposits in the Gulf of Venezuela, the Orinoco bitumen band, and the continental shelf, are an estimated 39,000 million barrels, of which 16,000 million have been discovered and 23,000 million are proved. But this estimate is not totally trustworthy because it was made by the oil companies themselves without being checked by any government authority.

New oil-bearing zones are being explored by companies which have signed service contracts: this is the new formula which has been brought in place of the concession, which by its very name stinks of colonialism. The deposits of heavy oils in the Orinoco bitumen band cannot yet be properly estimated at all.

And the many millions of hectares kept by the government as national reserves and now assigned to the Corporación Venezolana del Petróleo contain deposits of many millions of barrels of oil.

The distribution of gas for industrial and home consumption is now controlled by the Corporación Venezolana del Petróleo, as is a considerable proportion of the domestic market for petrol and other petroleum by-products.

One of Venezuela's serious difficulties, which successive governments have tried, and are still trying to remove, is the discrimination against us and in favour of our competitors by our principal customers, the United States. In 1959 the Eisenhower administration set up a mandatory system of oil import quotes, which does not help Venezuela at all, but ensures a better price for Canadian and Mexican oil (in the latter case it's a purely nominal concession, for Mexico exports very little to the US).

Venezuela certainly does not want to lower the price paid for Canadian oil; we simply want to be paid the same price for our oil, and for the US to give equal preference for the whole hemisphere without special privileges. When I was President, I went to Washington, in 1962. Being quite impervious to flattery, I didn't make the journey just for the pleasure of hearing the strains of the Venezuelan national anthem played on the steps of the White House. I went to discuss the unfair treatment of Venezuela and spent several hours with President Kennedy, giving him solid facts and figures to prove my case.

Kennedy promised me that he would satisfy Venezuela's wishes on this subject before he and I both left office. He was utterly frank with me and admitted that it was a 'devilishly difficult' problem to resolve because there was a whole network of vested interests at stake; but he assured me that Venezuela would get the fair treatment we demanded. His assassination in Dallas stopped this personal undertaking from being carried out. But our campaign, which was carried on with the same vigour under the governments of Raul Leoni and Rafael Caldera, will succeed and ensure equal preference for the whole hemisphere, to the benefit of oil exports from Latin America to the USA, which means basically from Venezuela to the USA.

Looking into the future

In Venezuela, as in all the oil countries, there will eventually come a time when the oil-wells run dry, for it is quite correct to say that oil is a typical non-renewable product of nature. For this reason, during the decades ahead until our oil runs out, we must develop a diversified economy for Venezuela, which will be able to weather the difficulties which will threaten the country when that happens. Quite a lot of obstacles have already been overcome in this direction. But, even so, the basic facts are that the country still depends on oil for one-fifth of its gross domestic product, two-thirds of its government income, and nine-tenths of its exports.

As a country, Venezuela is still revelling in its oil bonanza, with rulers and ruled alike squandering quantities of foreign exchange earned by oil. The amount of unproductive expenditure in Venezuela is comparable to the expensive display put on by Iran at the Festival of Persepolis. From 1959 to 1964, while I was in power, the national budget averaged 6,000 million bolívares. In 1972 Congress passed a budget which totalled 14,000 million bolívares. The national debt has followed the same geometrical progression as these huge increases in public expenditure, much of which is squandered. Venezuela desperately needs to put a check to this orgy of public spending, which is financed by an easily destroyed source of income. The first priority of government spending in Venezuela should be to provide more tangible services and free a large proportion of our people from the grip of poverty.

Mexico: Petróleos Mexicanos (PEMEX) is the state company which controls oil business here since 1938, the year in which the industry was nationalized by President Lázaro Cárdenas. In doing so he was protecting national sovereignty because the concession-holding companies had refused to accept a sentence handed down by the Mexican Supreme Court. Adequate compensation was paid to the companies which were exploiting the deposits under contract.

The 1970 annual report of PEMEX points out that the country's reserves amount to 5,000 million barrels, which, at the present rate of consumption, will last for 20 years. Production of oil and liquid gas averaged just under 900,000 barrels a day during that year. The value of the refined petroleum products imported by Mexico was higher than the value of all its oil exports: in 1970 it imported 13,500 barrels of petrol and diesel fuel a day, which cost 414 million Mexican pesos over the whole year (or 33·12 million dollars), as against a total value for oil exports of only 370 million Mexican pesos (29·6 million dollars).

PEMEX has run the industry without sticking too closely to the dictates of orthodox, intransigent nationalism. It has carried out its own exploration and exploitation programmes, alongside foreign companies which it has allowed in under the terms of service contracts. But nationalization in itself is not a panacea for all evils, as is clearly demonstrated by the fact that the state company hasn't managed to keep up with the needs of the domestic market, and its reserves will in fact give out in less than 20 years. The fact that the industry is under the control of the government of the producing country obviously makes very little difference when oil begins to fail and the inevitable exhaustion of the deposits takes place.

Argentina: The state oil company, Yacimientos Petroliferos Fiscales (YPF), was set up here more than four decades ago. Its aim is to make Argentina self-sufficient for fuel, which does seem to be a feasible proposition. By 1970 it was producing 425,000 barrels a day (in round figures). YPF hopes to get control of the domestic market for oil and its by-products; at present it controls 55 per cent, with Shell and Esso together taking 40 per cent, and the remaining 5 per cent shared out among small private companies.

Oil has been an explosive element in Argentinan politics. It is very difficult to make head or tail of the speculations and allegations on this subject, because of the aura of deep suspicion which surrounds all the activity of the oil companies. It has often been suggested, for example, that the *coups* which forcibly removed Presidents Juan Domingo Peron and Arturo Frondizi were set in motion as a result of the deals made by their governments with the foreign companies.

Colombia: Until recently this country produced sufficient oil to meet its own internal needs. But in the last few years production has gone down, and since May 1971 Colombia has had to import 15,000 barrels of crude oil a day from the Persian Gulf.

Colombia could make up for its lack of oil by buying oil from Venezuela on terms that would suit both countries. Other conciliatory agreements about oil are in sight: for example, Colombia and Venezuela seem likely to be about to set up a joint petrochemicals company.

These argreements would not only bring about a new level of economic cooperation between two neighbouring countries which are old friends. They would also make a big difference to the foreign policies of both countries. For they would help in the campaign, which the balanced and level-headed majorities of both countries are pursuing, to defuse the oil issue in the dispute between Colombia and Venezuela over their respective right to the continental shelf and to territorial waters: the boundaries are now being worked out in a realistic and civilized manner over the conference table.

Brazil: 'It's our oil' has been the war-cry of Brazilian nationalism. The deposits belong to the state and are run by a government company, PETROBRAS, which was set up during Getulio Vargas's first period in office. This company has signed various service contracts with foreign firms for exploration. But the efforts of the state company and of the contracting firms have all been fruitless. So far the country's oil has proved elusive and no significant deposits have been found anywhere in the enormous area of Brazil.

It is estimated that by 1980 this huge country, the economy of which is expanding rapidly, will be using 1·31 million barrels of oil a day, only one third of which will be locally produced. For this reason PETROBRAS has signed contracts with several oil-exporting countries to get oil in exchange for Brazilian products, under barter agreements. It has even been announced in the technical press that PETROBRAS is negotiating a contract to start explorations in Iraq in combination with the Iraqi state oil company, and is also trying to get concessions in Ecuador. Exactly the same conclusion can be drawn from this Brazilian story as from the Mexican story: nationalization alone does not help to produce oil in large quantities. It is self-evident that without considerable deposits, high production figures cannot be achieved.

(Note: In 1976 what I wrote here in 1967 has been confirmed by *The Economist,* which, on 3 July 1976, said: 'During the past few months Brazil's output from PETROBRAS has been declining. Production is now around 172,000 barrels a day. Brazil's consumption is approaching one million b/d and rising fast. Some of the onshore wells, which provide most of the domestic oil, are running dry . . . oil imports between January and April [1976] cost a third more than a year ago, and the year's bill looks like going up to 4 billion dollars from 3 billion.')

Peru: Though in colonial times Peru was overflowing with silver, it has not had good luck with its oil finds in modern times.

Peruvian oil production is less than 100,000 barrels a day, and large quantities of oil have to be imported to meet domestic demand. In 1969 alone Peru had to import 5 million tons of crude oil and petroleum by-products from Venezuela. In 1968 the government of General Juan Velasco Alvarado decreed the nationalization of the La Brea and Parinas concessions exploited by a subsidiary of Standard Oil of New Jersey; the legal validity of these concessions had long been disputed.

A state company, PETROPERU, runs these concessions now, and also the refinery at Talara. This state company has signed exploration contracts with several international companies, including Occidental and Texaco, to look for underwater deposits in the Pacific, and also to test the subsoil in the Amazon region, north of Loreto. And in November 1971 PETROPERU was able to announce that the first well drilled in this jungle region had struck oil.

Bolivia: Oil production is only 40,000 barrels a day, which is hardly sufficient to meet internal demand. However even this small amount, when linked with the hopes of finding rich deposits (in which Bolivia is no different from any other country), has sparked off many conflicts. There is a widely accepted theory that oil was the *éminence grise* behind the Chaco War between Bolivia and Paraguay, which was the longest and bloodiest struggle in the history of Latin America.

In 1938 the government of Colonel David Toro

nationalized oil, and set up a state company, called Yacimientos Petrolíferos Fiscales Bolivianos (YPFB). In 1952 this company controlled only 57 per cent of the national market. In 1955 the government of Víctor Paz Estenssoro drew up an Oil Code, the terms of which were meant to be favourable to Bolívia's national interests. Under this code agreements were signed with foreign firms, including the Bolivian Gulf Oil Company from the United States, which stepped up production and built a 450-kilometre pipeline from its oilfields in Santa Cruz to just outside La Paz, in order to link up with the YPFB pipeline to the Chilean port of Arica.

In 1970 the *de facto* government of General Alfredo Ovando Candia nationalized Gulf by decree. Russian experts were sent to Bolivia, and many people suggested that the Soviet government was about to take over Gulf's exploration and exploitation programmes in Bolivia. But it turned out to be pure talk; for the Soviet Union, stung by the cost of subsidizing its Cuban 'base', is hardly likely to make further prestige investments in Latin America.

Meanwhile government speeches full of revolutionary rhetoric do not help to put up production. For that working capital is required and that is just what the Bolivian government is not in a position to supply, since it depends on an annual gift of several million dollars from the US Treasury for its survival. Meanwhile governments come and go, following the jerky rhythms of constant military plots.

After General Ovando came General Juan José Torres, who produced still more fiery verbal pyrotechnics. Then Torres was kicked out by Colonel Hugo

Banzer, who came to power with the support of two civilian political organizations, the Falange Boliviana and the Movimiento Nacionalista Revolucionaria. Banzer's régime, which can hardly be described as a new hope, has still to make clear its policy on oil. Whatever happens, Bolivia's oil reserves do not at present appear to be of any size. Only the future can tell what large deposits are awaiting discovery.

Cuba: Cuba is a net importer of oil. In one of his verbose speeches Fidel Castro hinted that Russian and Rumanian experts had found traces of oil; but that was an empty boast. Cuba's economy is supplied with oil by tankers which come on the long haul from the Black Sea. Cuba depends so heavily on this Russian oil that, if the Russian tankers stopped going to Havana, within a few weeks the whole country would come to a standstill.

I can add a few details to the interesting story of how Cuba's absolute dependence on Russian oil came about. In early 1960 the oil companies, which were sending Venezuelan crude oil to be refined in Cuba, informed my government that the Cubans were refusing to pay for the oil they were receiving. Theodore Draper, in *Castro's Revolution, Myths and Realities*, says that Cuba owed the companies 76 million dollars. Draper was making a balanced and well-documented case against the 'Cuban model'. But even those enthusiastic supporters of the Cuban experiment, the US professors Robert Scheer and Maurice Zeitlin, in their book *Cuba: An American Tragedy*, admit that Castro's unpaid bills amounted to 50 million dollars. The oil companies demanded payment and threatened to cut off supplies. Castro's reply was to insist that they process Russian oil in their refineries. On 7 June 1960 the companies involved – Texaco, Standard Oil, and Shell – made it known that they were refusing to accept this demand; their refineries and the rest of their property were nationalized by decree and expropriated without any compensation being paid. In Venezuela we watched in consternation as we came near to losing the fifth most important customer for our oil.

We were very glad, therefore, when we heard that the Soviet ambassador in Mexico City, Mr Vazjkin, who was considered the Kremlin's top authority on Latin American affairs, wanted to come to Caracas to have talks with me and my Minister of Mines, about oil suppliers for Cuba. At that time Castro and his comrades had still not begun their campaign to provoke a civil war in Venezuela, and there were full diplomatic relations between Caracas and Havana.

In our talks with Vazjkin both sides were frank and stuck to concrete details. The ambassador explained quite openly that his country would have to make a great effort in order to keep Cuba supplied with fuel, and they would even be forced to hire tankers from Aristotle Onassis because they didn't have enough of their own. Vazjkin asked me: 'Could Venezuela go on selling oil to Cuba?' Perhaps he was surprised when I told him we could because he must have thought that we were tied hand and foot by the oil companies which operated in our country. I pointed out that we could take some of our royalties in kind, and, as a sovereign country, we could sell that oil to anyone who could pay for it. But I insisted that each shipment of oil to Cuba should be guaranteed by a deposit in advance in dollars or sterling in a bank in the US or Europe.

I added that, since our oil belongs to the Venezuelan people, no responsible ruler could dispose of it as if it was his own property. The ambassador, who spoke good Spanish, replied that he thought that my courteous but frank arguments were very reasonable ones. But the deal with Cuba went no further. Perhaps the Soviet Union considered the opportunity to keep Cuba on a tight rein worth the sacrifice of keeping her supplied with oil from the Black Sea. Certainly Cuba has become the Soviet satellite which is least able to break free from the yoke of Russian oil.

Chile: This country has a nationalized oil industry which it controls through a state company, ENAO. But it has had scant success in its eager search for oil. In 1970 Chile produced only 34,000 barrels a day, which was 6·9 per cent down on the previous year. The state refineries handle an average of 75,000 barrels a day, and the country has to import more than half its total crude oil requirements. President Salvador Allende has mentioned in a speech that Chile's oil imports use up a considerable proportion of the foreign exchange won by Chile's copper exports.

Ecuador: This Andean country is the latest winner of the oil lottery. In the Eastern region oil has been found in such quantity that at least eleven oilfields will be in production by 1972. 250,000 barrels a day will, when production begins, be piped across the Andes to the Pacific port of Esmeraldas. By a considerable stretch of the imagination, Ecuador can be seen as one of the future giants of the oil world.

A European oil journal has nicknamed Ecuador the Kuwait of Latin America and suggested that it has the same sort of potential. Let us hope that this potential is fulfilled and also that the government and people of Ecuador put this unexpected gift to good use by setting up more stable industries and bringing a higher standard of living to a people of great poverty. In countries which are suddenly flooded with unexpected wealth, which has arrived without any great effort being made, there is a strong temptation to spend lavish amounts of money, like a *nouveau riche*, on superficial improvements.

All the Latin American countries, even those which don't produce much oil, are anxious to get petro-chemical plants. Governments and businessmen alike are attracted to this idea because of the very wide range of articles that can now be produced using oil as a raw material.

Some conclusions

This paper has grown to a considerable length because it is impossible, or so it seemed to me, to analyse the recent history of oil in Latin America without putting it in the context of the world oil situation.

There are three main conclusions to be drawn from this attempt to put the present situation across in simple terms:

1 Today, and in the next few decades, oil is a vital factor for economic development and, as a result, for the improvement of living standards across the whole world.

2 The Third World countries, including those of Latin America, by setting up OPEC have shown the developing countries the best way to force the industrialized countries to pay a fair price for the raw materials they have to import.
3 The developing countries which are lucky enough to have oil should not be bewitched by their passing wealth; they should invest it wisely, so as to make lasting economic, social and cultural improvements.

Sources: *Petroleum Press Service* (London); *The Economist* (London); *Foreign Affairs* (New York); *Le Monde Diplomatique* (Paris); N. Gulbenkian, *Nous les Gulbenkian* (Stock, Paris, 1965); T. Penrose (ed.), *The International Petroleum Industry* (Allen & Unwin, London, 1968); T. Draper, *Castro's Revolution, Myths and Realities* (Praeger, New York, 1962); R. Scheer and M. Zeitlin, *Cuba: An American Tragedy* (Penguin, Harmondsworth, 1964); Press of Venezuela and other Latin American countries.

Rómulo Betancourt and One of his Deep Concerns as a Statesman: Venezuelanization of Oil

by Miguel de Los Santos Reyero

Former President Betancourt, in his speech in the Senate, made few references to the vital part he has played in the long sequence of events which has led up to the nationalization of the oil industry, despite the indisputable fact that he is the Venezuelan who has devoted most time and energy to studying and popularizing everything connected to this subject.

Forty years ago . . .
Since he returned to Venezuela, in 1936, from his first period in exile, he has been, without a shadow of doubt, the man who has contributed most to the formation of a nationalistic awareness with the goal of

state control of our basic wealth. No other Venezuelan has maintained a comparably high level of both spoken and written discussion of the oil question. In public meetings, lectures, pamphlets and books over the decades he has protested against all the negative effects on Venezuela of the exploitation of the subsoil by the concession-holding companies, and has pointed out possible ways to put an end to their abuses. In both his periods of office as head of state, his governments took fundamental steps in oil affairs which paved the way for the nationalization of that industry. While in exile in the last years of the Gómez régime, he began a book on the Venezuelan oil industry, which he completed while in hiding and in 1938 had ready for publication. In 1937 one of its chapters was published in Caracas under the title of 'A republic for sale' with a cover drawn by 'Medo' (a pseudonym for the great cartoonist and dermatologist Dr Mariano Medina Febres). The printers who printed the pamphlet had to pay a fine imposed by the governor of the Federal District of Caracas, and the man who took the manuscript to the printers was held by the police for two weeks. Subsequently the manuscript was sent to Mexico to be published by a group of fellow members of the Partido Democrático Nacional. But they didn't manage to bring it out because they didn't have funds to pay for it to be printed. So Betancourt was never able to publish his first book on oil.

Political activity while in hiding in Venezuela

Betancourt went underground and was active for three years from 1937 to 1940 with an enthusiastic young group laying the foundations of his party Acción

Democrática; and at the same time he was writing an unsigned daily article in the Economics and Finance section of the democratic Caracas newspaper *Ahora*, in which he very often dealt with issues connected with the oil industry. This is borne out by the fact that one-third of the book of articles reprinted from *Ahora*, which he published in Chile when in exile there in 1940, under the title 'Problemas Venezolanos', were grouped in one chapter under the heading 'Oil, problems and possibilities'.

During these three years of underground political activity Betancourt did not just write press articles on the oil question. He also made his nationalist influence felt in National Congress through intermediaries. In his book *Venezuela, Política y Petróleo* (3rd edition, Editoral Senderos, Caracas, 1969) he writes about this anonymous nationalist campaign, which he carried on without any publicity at all, and without trying to get popular appeal either for his party or for himself. On page 140 he says:

> National Congress in 1939, when we already had a few PDN deputies, also certainly set the cat among the pigeons for the managers of the Venezuelan sections of the oil companies and their Venezuelan collaborators.
>
> The Upper and Lower Chambers met in a joint session to hear the report of the Hydrocarbons Committee, which had been able to study the chapter of the 1938 Ministry of Development Report concerning oil. The report of the Committee gave a crystal clear picture of the way in which Venezuela was being treated as if it was still

a colony. One of its key paragraphs runs as follows: 'This Sub-Committee's attention has been drawn above all by the fact that, while the government's revenues from oil royalties and taxes in 1938 amounted to 110 million bolívares, the value of the oil companies' exemption from import duties in the same period totalled 95 million bolívares.' (*Diario de Debates de la Cámara de Diputados* No. 27, 31 May 1937, p. 4)

In other words, the real value of the Venezuelan government's share in the fabulous wealth of our subsoil was only 15 million Bs in 1938, because the 'legal' trickery of the exemption from customs dues almost cancelled out the small quantity of taxes collected. Congress reached the further conclusion that this sum was 'all the more scandalous when compared with the enormous profits and cash dividends announced by the oil companies'.

That same Congress, when voting on the Development Ministry Report, agreed unanimously on the following judgement, which summed up the nation's hopes: 'We urge the Executive to endeavour by all possible legal methods to increase the Venezuelan people's small share in the oil extracted from our subsoil, as revealed by the Minister of Development in his Report.' These were arguments, ideas and even words used by the PDN, and now adopted by an official body over which, as a party, we exercised only a very weak influence through our meagre number of deputies. We made up for our weakness with cunning. From underground, we both fought and intrigued. Our intrigues were legitimate because we did not aim to win popularity for a political group or its leaders, but to defend the national interest. We used unlikely channels to get our ideas to the public, by subtly turning official government spokesmen into active spokesmen for an underground party. Some of our political enemies in Congress never realized who was drawing up the motions they were putting forward in the Chamber or in the Senate, nor that several reports by various Congressional committees were not draughted in the Congress building, but in a backstreet proletarian home, on the silent typewriter of an outlaw wanted by the political police.'

That outlaw was Rómulo Betancourt. This was the way he smuggled his nationalist ideas into Congress, not only via the very few deputies who belonged to the underground PDN. In his task of defending the nation's interests he was helped by a close friend without party loyalties, who on several occasions gave Betancourt refuge in his house, the Lara congressman Dr Julio Alvarado Silva, who spent years in a cell of the castle at Puerto Cabello during Gómez's reign of terror.

In 1941 Betancourt returned from his exile in Chile and in that year the old PDN, after changing its name to Acción Democrática to overcome government resistance, became legal.

Betancourt as Secretary General and Founder of Acción Democrática

Between 1941 and 1945 Betancourt travelled all over the country, sometimes accompanied by other Acción Democrática leaders, organizing party cells even in the smallest towns and villages of our wide country. In his speeches and lectures all his arguments centred on the oil question. In 1943 his speech at the so-called Concentración de los Caobos provided the only jarring note among paeans of praise for General Medina Angarita, who was President at the time, and his proposed reform of the legal relationship between the state and the industry. At that moment nobody in the country outside the highest government circles knew exactly what terms the government had in mind for this reform which was of such great importance to the country. The bill for a Law of Hydrocarbons finally reached Congress in 1943, in a hurry and almost by stealth, and Betancourt again used newspaper articles and public speeches to give generous support to the positive aspects of the bill, and also to denounce the harmful effects it would have on our country's interests, by producing indisputable facts and figures, and without putting on any demagogic airs.

Former President Betancourt did mention in his speech in the Senate the assistance he gave to his old friend and colleague Dr Juan Pablo Pérez Alfonzo (who has acknowledged this help on several occasions) in preparing the speech to justify the latter's abstention from the vote on that bill in the Chamber. Betancourt also wrote the justification for the abstention of the Minoría Unificada, the parliamentary group formed by deputies of Acción Democrática and other political independants. In the original typewritten manuscript of the Minoría Unificada speech appear corrections in Betancourt's own handwriting, and it is, of course, couched in his inimitable style. Both these important documents, which are invaluable historical records now that the oil industry has been nationalized, are to be found in the archives of National Congress.

Betancourt as President of the Junta Revolucionaria de Gobierno, with Juan Pablo Pérez Alfonzo as his Minister

After the memorable events of 18 October 1946 Betancourt became President of the Republic for the first time. He called on Dr Pérez Alfonzo to fill the Ministry of Development, which then included oil affairs in its sphere of reference. These two men were linked by close party ties, by personal friendship and by a common deep concern about the practices of the companies which produced, refined and marketed the hydrocarbons extracted from Venezuela's subsoil.

Former President Betancourt gave a brief summary in his speech in the Senate of the nationalist measures taken in the first three years of Acción Democrática government (1945–48) in relation to the oil industry. For nine months of that period Rómulo Gallegos was President, until overthrown by a conspiracy headed by his own Minister of Defence, Colonel Carlos Delgado Chalbaud, on 24 November 1948.

The first collective contract between the oil companies and their workers

Betancourt also gave a very brief account of the first collective contract between the oil companies and their employees, signed in 1946. No previous agreement of this kind between workers and management had been reached. The situation of the oil workers and their families in terms of wages, social services and general living conditions was demoralizing and had reached dramatic extremes.

On 30 May 1946 this collective contract was signed by representatives of the companies and of the workers. Luis Tovar, who founded and led the Oil-workers' Federation of Venezuela, stated at that time that in the previous eight months they had got concessions which they had been unable to achieve in the thirty years during which the companies had operated in Venezuela. Tovar was a trade unionist active as a member of Acción Democrática; he became National Senator and died one year ago while serving as President of the World Federation of Oil and Atomic Industry Workers, an honour which he fully deserved, for his intelligence, experience and honesty. In one passage of his 1946 message he said:

> We have won a great victory for the oil proletariat. In less than eight months since the October Revolution we have done what we couldn't do in the thirty years that the oil companies have been exploiting our country. In 1936, when we undertook the glorious oil strike, and demonstrated the high morale of the working class, we only got a token wage rise of one bolívar. We then tried again with the First Congress of Oil Workers and didn't achieve anything. Now we have got both our economic and social demands satisfied. I am sure that all my followers in the oilfields will be celebrating the news of this agreement, first of all because of the victory we have won, and secondly for love of our country, for they realized that a strike problem would harm the Venezuelan economy and endanger

> our revolution: we have a right to call it ours because we take part in it. (*El País*, Caracas, 31 May 1946.)

Betancourt's third exile and the book 'Venezuela, Política y Petróleo'

During his third period of exile (which lasted from 1949 to 1958), Betancourt continued to keep himself well-informed about developments in both the Venezuelan and the world oil situation.

His book *Venezuela, Política y Petróleo* was the product of his practical experience in government and his ceaseless study of the many facets of the industry. It was published in 1955 by Mexico City's Fondo de Cultura Económica, the most important publishing house in Latin America, with the approval of Licenciado Victor Urquidi, one of Latin America's most influential economists, who is at present head of the Colegio de Mexico, the most respected centre of academic research in Mexico. After approving the publication of the book, Urquidi wrote to Betancourt from Mexico City on 12 April 1955 in the following terms:

> I believe you have written a valuable book not only because of what it reveals about the contemporary history of your country, and because of its political significance and literary value, but also, from the point of view of what should concern all Latin American economists, for its crystal clear account of what is involved in carrying out a programme of economic (and social) development. It teaches the need for clear, interconnected goals, the fundamental

importance of institutional reform, of education, health, labour organization, taxation, etc., and the need to get things moving even without good statistical and technical information. What is most striking is the instinctive good sense of the economic programmes, despite the lack of basic information which many economists would consider indispensable. Of course it must be remembered that Venezuela's position was unique in that, in the short term, and still for some years to come, no serious financial or monetary problems could arise. Under these circumstances development is relatively easy to stimulate. But you describe brilliantly the difficulties encountered putting these projects into practice and coordinating them, and provide an object lesson for the theoretically-minded.

There is little I can add about the book as a whole. It is so complete and rounded that one ends up with a very full knowledge of the problems of your country and the influences, both positive and negative, generated by its main export. The historical account and the section about ploughing back the benefits of oil into society I found fascinating. My attention never wandered throughout the book, not even in the parts with detailed facts and figures, because they are so well chosen that they effortlessly convey what is intended. For what my modest judgement is worth, I believe that you have written a book of great importance for Latin America.

'Venezuela, Política y Petróleo' on native soil

Betancourt has been accused more than once, by ill-informed writers and political opponents, of not publishing the second edition of *Venezuela, Política y Petróleo* after he returned from his third period of exile in 1958 for fear of reprisals from the oil companies. Some writers have gone further and have claimed to find important differences in facts and figures between the first edition (1956) and the three subsequent editions (1966, 1967 and 1969).

Betancourt has dealt very well with the first of these false charges. His old friend Ricardo Montilla had arranged for a new edition to be brought out by Las Novedades, a publishing house then owned by Miguel Angel Capriles. But Betancourt in the end decided not to publish a new edition of his book at that moment. He explains why on page 12 of the second edition of *Venezuela, Política y Petróleo* (Editorial Senderos, Caracas, 1967), where he says:

> I had two reasons for doing this. It is a polemical book, which analyses and passes judgement on contemporary politics in this country. Because I was sure that I was going to win the elections which were then about to take place, I didn't think it was the right moment to publish such a controversial book. Secondly, I considered it improper for a head of state (and I was certain that I would soon hold that office) to be in a position to use his political office to promote a book he had published.

Ricardo Montilla adds that Betancourt, after becoming head of state, signed a 'cheque for 20,000 bolívares for the publisher Capriles to cancel an advance payment he had made for the book. Montilla himself gave the money to Capriles.'

It is quite wrong to accuse Betancourt of making substantial changes and cuts before the book was reprinted. Anyone who compares the 887 pages of the Mexican edition with the 943 pages of the third edition (Editorial Senderos, Caracas, 1969) can see that. As he himself has said, Betancourt was concerned to write contemporary history and not to harm the reputations of individual Venezuelans whom he considered guilty of mistakes but not of crimes against the country, and he did remove some words by which people whom he respected had been injured.

As proof of Betancourt's unceasing interest in everything to do with the oil industry, we have the long epilogue to the second, third and fourth editions of the book, written during his voluntary retirement at Berne, in Switzerland. This is an analysis of the changing oil situation in Venezuela and the whole world, up to December 1967. The book is still selling well in this country's bookshops because the public's interest in the subject is constantly being renewed, and also because universities and secondary schools have adopted it as a textbook on the oil question for students.

In Berne Betancourt also wrote a long article for *Vision*, in 1971, which is reprinted in this book under the title, 'Oil, Ruler of the World'.

The Junta de Gobierno's 1959 decree, when Betancourt was President-Elect

Former President Betancourt has recently explained why he made no mention in his speech to the Senate of the decree promulgated by the Junta de Gobierno under Dr Edgar Sanabria in 1959 which raised taxation on the oil companies' profits. Betancourt said that he reluctantly left this out because he was determined to limit the length of his speech in Congress to two hours, as he has since made clear. He pointed out that he had, of course, made favourable comments about this decree in the lecture he gave to one thousand professional men the night before he gave his speech in Congress.

This shows that Betancourt has always put the country's basic interests above his personal feelings. On page 945 of *Venezuela, Política y Petróleo* (1969 edition) he speaks of 'the slightly unorthodox practice of promulgating this decree, when there was already a president-elect'. What happened was that, although he had got 50 per cent of the vote and was effectively already President of the Republic. Betancourt was not officially consulted before the decree was promulgated. To clear up his words about 'slightly unorthodox practice' we might add that Betancourt had used as the keynote of his election campaign the need to increase taxation on the enormous profits made by the companies under the dictatorship of Pérez Jiménez, and of course by then it was known that the party which he himself had founded and which had chosen him as candidate for the presidency had obtained a comfortable majority in both houses of Congress.

The oil question was the focus of his presidential campaign

The best examples of Betancourt's strong emphasis on the oil question in his presidential campaign are to be found in his lectures to businessmen, collected in book form in *Rómulo Betancourt, posición y doctrina* (Editorial Cordillera, Caracas, 1958). On page 83 of the first edition of this book (which was quickly reprinted) begins a verbatim record of his speech to the Maracaibo Chamber of Commerce. This lengthy excerpt needs no comment:

> I have said that we are not getting a proper share in our fantastic oil wealth. We can see this easily by analysing the figures produced by the Ministries of Finance, and Mines and Hydrocarbons, and by the Venezuelan Central Bank. The profits of the oil companies in our country amounted to 1,201 million bolívares in 1951, and six years later, in 1957, had gone up to 2,765 million, an increase of 125 per cent. The 1951 profits worked out as 23·35 per cent of the capital invested, which is a very high rate for an industry which is normally considered on a par with public utilities. But by 1957 this percentage had gone up to 32·43 per cent, which is an excessive rate for any kind of investment, and even more unreasonable in view of the fact that the oil industry has recovered the capital it invested several times in thirty years.
>
> These figures are based on data given by the companies themselves. Under the régime in power until January of this year, which had no sense of responsibility to the nation, it has been quite impossible to analyse two additional factors; first of all the strange and unexplained fluctuations of the price of our oil below the normal level in the world market; and secondly the profits made by the subsidiaries and intermediary companies linked to the oil-drilling companies, above all the refining companies. There is some evidence which suggests what

surprising discoveries would be made if there was a proper investigation of the additional income which the companies get from processing the oil they produce in our country, for instance from their refineries. In 1954 the US Department of Trade investigated the profits on US capital invested in Latin America and published its findings. They discovered that US investments in the Dutch islands of the Caribbean produced enormous dividends. Those investments, in plain language, are those of the Creole Petroleum Corporation in their refinery on Aruba. In 1954 their total investment was 296 million dollars, and their profits were 104 million dollars. Their investments worth 296 million dollars produced a profit of 104 million dollars, in other words getting on for half the capital invested.

In 1957 the Venezuelan Central Bank, with justifiable caution in view of the political situation of the country at that time, produced the figure of 2,388 million bolívares for profits on foreign investments in Venezuela, and drew attention to what the United Nations Economic Commission for Latin America (ECLA) had said in its 1956 Report, namely that half the total profits on foreign investments in Latin America came from Venezuela.

The situation demands a reform of the relationship between the state and the industry to increase the income from what is just about the only source of government revenue. In addition to reforming the tax situation, we must start a farsighted and responsible oil policy.

Former President Betancourt, not being given to

recriminations, did not protest at all about what was to put it kindly a rather clumsy move by the provisional government. The 1959 decree favoured the nation's interests and it was Betancourt's government's job to put it into practice, as he announced to the nation in his first Message to National Congress on 13 February 1959. This is what he said in his official speech about the unreasonable profits made by the financial groups exploiting our oil:

> The provisional government's reform of the Income Tax Law will not be modified in the present economic situation to allow further tax increases, but it will prevent the difficulties facing the Nation's Treasury from reaching alarming proportions. No one can deny that we have scarcely found the nation's coffers overflowing; they are almost empty as a result of the policies carried on under the dictatorship, which squandered what income remained unembezzled.

Betancourt condemns the US government's attacks on OPEC

When the preceding pages had already been completed as an epilogue to this book, Betancourt agreed to write for us an analysis of the hostile position taken up by the governments of the industrialized countries against the rise of oil prices and against the very existence of OPEC. We publish his opinions in full beneath and they confirm once again that Rómulo Betancourt is still one of the best-informed men in this country in relation to oil problems at both the national and the international level, with a very clearly-defined point of view:

The rise in oil prices and the formation of OPEC as an agreement between small producers to defend themselves against being plundered by the multinational companies and by the governments of the capitalist world, have been viewed not kindly, but with hostility, by the governments of the West. They would have liked to be able to go on beating down the price asked by the oil-producing countries of Latin America, North Africa, the Middle East and parts of Asia, and buying the vital raw materials they possess (which cannot be replaced once it leaves the ground) at a minimal price. Then they would have gone on selling them in return for capital goods, such as machinery and other inputs, needed for development, at constantly rising prices. The present US administration has been the most aggressive of all in this respect. President Ford, his Secretary of State Henry Kissinger, and the abrasive Mr Simon, the Secretary of the Treasury, are those who have the doubtful honour of using the strongest language about OPEC and its member countries. Inside the United States, where dissent is not silenced by totalitarian restraints, important figures, with reputations in scientific research or in government, have produced calm and convincing criticism of these angry tirades. Robert McNamara, who played a very important part in the Kennedy and Johnson administrations and is now President of the World Bank, has gone on record openly contradicting the unbalanced opinions of the men in power at Washington. He produced figures which prove that inflation was already hitting Western Europe and the USA before the prices of liquid fuels went up, and taught a simple lesson to those who use the hypocritical argument that it is the poor countries of the Third World which they are really trying to defend against the effects of those price rises. McNamara reminded the rulers of the industrialized countries that they have never fulfilled the solemn obligation they undertook through the specialized organizations of the United Nations to spend of their GDP on development programmes for the Third World. He went on to show, and I summarize his argument without using his exact words, that the OPEC countries are helping to relieve that abject poverty of the Third World by making large loans in petro-dollars on liberal terms.

A better example of expert opinion in the United States about that country's own internal energy resources, and the price of imported fuel, was the full centre-page article in *The New York Times* of 19 March 1975, published by the specialist writer Victor K. McElheny. President Ford had readopted President Nixon's so-called Independence Programme in a slightly less optimistic form and had produced a plan by which the USA would become self-sufficient in energy in a relatively short time. He recalled the increases in production made during the Second World War and proclaimed: 'We did it then. We can do it now.'

President Ford's reference to the great efforts to push oil production up by leaps and bounds during the Second World War can be backed up with figures from an official US government memorandum of the time published in *Foreign Relations of the United States, 1948* (United States Government Printing Office, Washington, 1972, p. 245): 'The result of World War II was a tremendous strain on the oil reserves of the US. The reserves we held at the beginning of the war were used up at the rate of over one million barrels a day. During the war the average production of crude oil in the US increased from 3,606,157 barrels a day to 4,871,099. Since the war, world demand for oil has reached the unprecedented level of 9 million barrels a day of which 5,600,000, or 62 per cent, are supplied from the reserves of the USA.' This source also gives an estimate of the reserves of the Western hemisphere in barrels in 1948. I will only reproduce the figures for the United States and Venezuela. The US had known reserves of 4,577 million barrels, and its daily production had risen to 1,616,600 barrels. The figures for Venezuela were reserves of 5,500 million, and daily production 622,000 barrels. In 1975 the situation of the oil resources of the US is very different from this 1948 estimate, as Victor K. McElheny said in his well-documented article:

> Energy experts think not only that the oil and gas reserves of the USA are much lower than government estimates, which are based on geological studies, but also that it will take from 5 to 8 years to bring those reserves into production. The National Academy of Sciences recently echoed the pessimistic views often expressed by the geologist King Hubbert. Hubbert's estimate makes the total reserves of the US around 170 billion barrels, as compared with estimates based on geological studies of 600 billion. Of those 170

billion, 100 have already been discovered. Of the remaining 70 about 6 could be brought into production by 1985 if federal government projections are correct.

On the basis of detailed evidence from experts and scientists, the *New York Times* writer gives the government's plans no chance of success. The Apollo programme did succeed after a decade in putting North Americans on the moon, but the Ford Plan seems destined not to have any such success. The government's aim to supply internally all the country's requirements for its gigantic industrial complex and its powerful war machine is a very doubtful proposition indeed.

In order to build large numbers of new atomic energy plants for peaceful purposes, to use liquid coal to produce petrol, to exploit bituminous schists, to harness solar energy, and to make full use of Alaska's oil reserves, the USA would have to use too much capital and building material to make this plan a technically feasible proposition. McElheny summed up as follows the consensus of all the experts he consulted:

> The enormous construction effort that President Ford wants would cost 400,000 million dollars. It could require from 140 to 160 million tons of steel, made into hundreds of steam turbines for electricity and hauling cables for coal mining, thousands of coal-cutting machines for underground mines, dozens of oil and gas storage tanks for Alaska, between 8,000 and 10,000 rail-

way engines, and as many as 260,000 hopper wagons for transporting coal, plus thousands of miles of steel piping for oil wells, refineries and oil and gas pipelines.

As for the price of crude oils, the Ford Plan is based on an estimate of 11 dollars a barrel in the next few years. But this estimate was challenged by two very respected oil experts, Doctors Abelson and Linder, who incidentally defended the oil-producing countries. Dr Linder pointed out that 'the exporting countries are selling their product at a price below the cost of replacing it. In fact the OPEC countries have been very generous to us.' What is more Dr Abelson, in a lecture at Washington University, said that what happened to oil prices during the 1973–4 embargo (put on by the producing countries in the Arab world) showed that 'a price of 20 dollars per barrel would be a more realistic upper limit than the government estimate of between 7 and 11 dollars'. He said finally that the Arabs – and also the Venezuelans, I would add – possess limited resources of oil which cannot be replaced, and asked: 'Even if these countries could bring the price down again, why should they do so?'

The dilemma forced by the USA as a result of their energy requirements and the present prices can be summed up in the following terms:

1 There is no serious evidence to support the theory that the US will be able to stop importing combustible fuels.

2 The producing countries, now that they are united in the common defence of their interests by an international body, are no longer prepared to allow themselves to be exploited by the industrialized countries.

These two facts are perfectly obvious and clearly suggest that the verbal pyrotechnics inspired by Washington against the countries which produce the world's oil not only are unjustified but should be totally condemned. At least from the point of view of this sub-continent's oil-producing countries, this policy by the US seems a strange way of implementing the 'new dialogue' with Latin America, on which Washington has put great emphasis.

But the aggressive tone used by the present US administration against the OPEC countries could not lead in the present world situation to the extreme solution of armed invasion to occupy countries which have thrown off the yoke of semi-colonialism.

On 5 September 1975 United Press sent out a long report from Washington signed by John J. Walte, which said: 'Venezuela and Nigeria, the principal non-Arab countries in OPEC, would be easy targets in the case of a US military invasion of their rich oilfields according to a "feasibility study" prepared for US Congress.' But this summary of an analysis commissioned by a sub-committee of the Foreign Relations Committee of the House of Representatives sagely concluded: 'On the other hand the same study warns that the disadvantages stemming from political and military operations of this kind would be so great as to make this policy a totally counter-productive one for the USA.' An important factor, in the case of a punitive expedition against Venezuela, would be the position of loud and aggressive hostility to the United States which all of Latin America would adopt. I might also slip in an emphatic mention of another very important factor. I believe that nobody of any importance in the decision-making process of the US government has even considered the possibility of using that country's powerful war machine to force Venezuela to go on making what amounts to almost a free gift of the non-renewable resources of her subsoil to the US. Suppose a North American head of state were to go mad and order the use of arms against Venezuela, it would not be an easy policy to carry out. For a start he would find that sane public opinion in the US would put up strong resistance against such an unwise venture. Secondly, what is far more important, we Venezuelans would fight, men and women alike. At the time of our wars of Independence men and women of this country – half the population at that time – died on battlefields not just in Venezuela but also in other countries which we were helping to achieve their independence. The fighting blood of the generation which founded this Republic still flows in the veins of the 12 million men and women of this country.

Let me make one last observation. If the strong speeches and statements emanating from Washington against OPEC do not pay off, and the threat of military intervention is laid aside, why shouldn't the producing countries meet the consuming countries round a conference table. The president of

France, Giscard d'Estaing, is still working hard to bring about the meeting he suggested on a most sensible and rational basis.

(*Note:* These remarks about the US government's hostility to OPEC were written on 10 September 1975. In 1976 there have been distinct signs of a change in the US position. They have made a definite change of tack, bowing down to what the British would call 'the obstinacy of the facts'. On 12 July 1976 *The New York Times* published an article by its leading economic correspondent, Clayde H. Farnsworth, on this change of tack. The article quotes a recent speech by Secretary of State Henry Kissinger who admits that the industrial countries' purchases from the member countries of OPEC will rise in the next decade from 27 million barrels a day to 37 million. This is in line with the analysis published by the International Monetary Fund's oil expert Mariano Gurfinkel, who points out that production by OPEC members could rise by 12 per cent in 1976 over the figures for the difficult year 1975. In the first three months of 1976 world oil production has gone up by 2·25 million barrels a day, which is 4·4 per cent above the figures for the same period of 1975. This rise in oil production goes hand in hand with the intensification of industrial production in the developed countries of the West as they recover from the recent recession. In the first four months of 1976, domestic consumption of oil in the USA went up by 3·5 per cent, in Japan by 4·7 per cent, and in the top four countries of Europe by 5 per cent. These eloquent figures serve to justify the conclusion Farnsworth reaches in his *New York Times* article, which is: 'After months during which the United States focused its international diplomacy on the goal of reducing the influence of OPEC, the government has realized that this organization holds a very strong hand, and is therefore developing a more realistic policy.') (Note dated 14 July 1976.)

Betancourt's deep satisfaction at the realization of these goals

This has been a lengthy narrative, but it will have become apparent how necessary it was to fill in the background to the speech delivered by former President Betancourt, and to evoke the almost pig-headed determination of a Venezuelan, who, along with many others, has helped to pave the way for an event that will go down in the history books of Venezuela and of all Latin America: the ratification and signature by President Carlos Andres Pérez of the Organic Law which reserves for the State the production and marketing of hydrocarbons.

This law transferring the full control of Venezuela's basic wealth out of foreign hands was signed and sealed on 29 August 1975, in a ceremony which, fittingly, took place in the Oval Room, where the original declaration of Venezuelan Independence is kept, with the signatures of our founding fathers, dated 5 July 1811.

While President Pérez was giving his eloquent and enthusiastic speech to the country, the television cameras on several occasions focused on former President Betancourt, and former President Doctor Rafael

Caldera, and the former Presidents of Juntas de Gobierno Rear-Admiral (retired) Wolfgang Larrazabal and Dr Edgar Sanabria.

The happiness and joy in Betancourt's face were plain for all to see; for he had played a leading rôle in the struggles through which Venezuela has had to go to recover her oil wealth.

But the country faces two challenges, the former President said:

> First, to administer efficiently and honestly the nationalized oil industry. Secondly to ensure that the profits derived from the industry, now that they are fully controlled by the government, should be used in a planned attempt to develop the national economy so that Venezuela no longer depends on her mineral resources as her sole source of wealth, and also to bring about a more equitable distribution of wealth, which will give all sectors of the community, above all the lowest-paid workers, access to welfare services and education.

We are certain that the country can be confident that Rómulo Betancourt, without any trace of personal political ambition, will make a truly Venezuelan contribution to the full success of the nationalistic venture which he, almost above all others, has helped bring into being.

Caracas, 10 September 1975

Documents

Abstention of Doctor Juan Pablo Pérez Alfonzo

From the vote of the Chamber of Deputies' Development Committee's report on the bill for the Law of Hydrocarbons, on its second reading.

I, Juan Pablo Pérez Alfonzo, believe that, in general terms, the positive elements in this bill can be divided into two groups: technical and legal improvements, and changes that will benefit the national economy. On the one hand it simplifies the relationship between the concession-holders and the state, and defines more clearly the ways in which the state can intervene in the industry. In economic terms the following reform can be approved without any hesitation:

a) the considerable overall rise in taxes (even though some taxes which the concession-holders have paid up to now are abolished),
b) the abolition of customs exemptions,
c) the control of oil transport,
d) the clause which forces the companies to keep their accounts in Venezuela,

e) the abolition of the unjust differential taxes assessed according to the location lots distributed among the different concessions,
f) the projected agreements about refineries.

All these positive moves will undoubtedly bring the resolution of the oil problem closer. But when the government maintains that these gains mean that the past is wiped away completely, and 'all previous failings and any lawsuits and claims arising from them are done away with', it is making a quite unjustified boast. For it is a notorious fact that the oil companies have plundered the wealth of the Venezuelan people despite being aware of what this country's needs are. The companies have taken advantage of the weaknesses of those who have represented this country, whether legally or illegally, and have made illicit profits and left a trail of damage which cannot simply be cancelled out by a clause in a law: there is no legal sleight of hand which can transform injustice into justice at the stroke of a pen.

As a rough estimate it can be said that the total volume of oil produced from Venezuelan wells up to now is 2,500 million barrels worth at an average selling price of one dollar a barrel, 2,500 million dollars. If we take 50 cents as the average cost of production of a barrel including taxes paid in Venezuela and amortization and interest payments on capital (as proposed by the late Minister of Development, Dr Manuel Egaña, in his report to Congress last year, on the basis of figures prepared for the US Tariff Commission) we can see how justified is the present wave of protests against the exorbitant profits of the companies. The profits total 1,250 million dollars (or 3,800 million bolívares), which is more than the estimated value of all the capital investments in Venezuela, in agriculture, stock-raising, and industry, including mining and the oil industry itself. It is hardly reasonable to expect the Venezuelan people to allow the companies to reap such massive profits which strike at the very basis of the country's wealth and at its fundamental principles, justice and social welfare.

The total purification of the Venezuelan oil industry, its ritual cleansing, will remain impossible until the companies have paid adequate financial compensation to our country. Any other (apparent) solution is in reality nothing more than an attempt to postpone the hour of judgement for the companies, unless the government is considered to be able to wield supernatural powers. In this context it is well worth recalling the words of Charles Hughes, a president of the US Supreme Court, on a similar subject:

> It should be openly recognized that many of the evils about which we complain spring from the laws themselves, from carelessly granted privileges, from opportunities given to private interests to make a profit at the expense of the people, without any safeguard for the public interest which would provide means of controlling the companies which enjoy these concessions.
>
> Wherever the law gives an unfair advantage, wherever it fails to protect the public interest with suitable controls, wherever powers granted by the state are wielded against the state, the authority of the state not only should but must be upheld.

This bill creates a special privilege, by giving up all the nation's other possible claims against the companies. But this will be nothing more than a legal formula without the full support of the Venezuelan people. Such a settlement could never be accepted by our people because its basis is totally unjust. The Venezuelan people has been deeply affected by its losses at the hands of the oil companies; it cannot accept that the more tightly-controlled relationship between the state and the holders of the oil concessions will apply only to the future and not to the past. It is quite unreasonable to think that justice in the future can somehow make up for damage done in the past. The claims behind every single positive aspect of this law are ones which might quite justifiably be made in any country, and in some cases the claims do not go far enough. The companies should be forced to pay a lump sum in compensation. That is quite a feasible demand to make, because the companies have already paid compensation in connexion with the settlement of certain isolated infringements by some companies since 1935. Last year the state got 30 million bolívares out of one of these legal settlements, which I challenged because I thought it had been reached without a deep enough examination of the issues involved, and without a high enough figure for the compensation. Since there are concessions which are many times as large as Lot 5 of the López Rodriguez concession which was at issue then, it is quite safe to say that the total purification of the industry ought to produce a large sum, which, if well invested, would make up in a tangible way for the injustices suffered by the Venezuelan people.

In short, I believe that this bill contains many good points, on which everyone is agreed, and that it gives a reasonable basis for the concession-holders to operate on, though in many cases our country could fairly get still better terms. But for this reason I do not believe that this new basis is at the same time an adequate substitute for payments to compensate for the illegal activities of the companies in the past.

I have already pointed out that, as regards the future, this bill sets up conditions which are reasonable from the point of view of the holders of the concessions, and, from our country's point of view, are much better than those which were in force before; but, on many issues, it only goes part of the way towards a just and reasonable solution.

It is a distinct advantage from an administrative point of view to get all the oil concessions on to the same legal basis. But a more effective reform, which would have been by no means impossible to carry out, would have been to make all the concession-holders form a single large company, which would then be more directly under the control of the government; a settlement along these lines was proposed by these same oil companies to the Mexican government in 1938 through their joint representative, Donald R. Richberg (see the Mexican government publication, *La Verdad sorbe la expropiación de los bienes de las empresas petroleras,* 1940). On these lines the simplification of the concession system would work better, and this bill would be more likely to achieve what it sets out to do in terms of establishing overall control by the state of an industry which draws on publicly owned resources, and also in terms of putting an end to the evils of competition.

As the preamble to this bill states, and as the Minister of Development has pointed out, it is perfectly fair to put up taxes in order that the state may get a higher share in the profits made from our oil than the companies. Here are the premises on which the bill has been based, which reveal its fairness even more clearly: a barrel of Venezuelan oil sells for 81 cents, and it costs 40 cents to produce, the profits of 41 cents will be divided between the government and the companies, with the former getting 24½ cents and the latter 16½ (figures taken from the lecture given by Dr Edmundo Luongo Cabello, a Development Ministry official). This figure of 16½ cents a barrel seems a fair return on investments totalling, say, 400 million dollars in this country: annual production averages 200 million barrels, which means profits of 33 million dollars for the companies, or 8 per cent a year for the capital invested.

The dictum of the English judge, Lord Hale, is universally accepted today: 'When a private property is affected with a public interest it ceases to be *juris privati* only.' In Venezuela there can be no doubt that the oil industry is of vital interest to the nation, not only because it is defined as such in several laws passed by Congress, but also because it is in actual fact the mainstay of the nation's livelihood and its whole economy. So the private companies are perfectly aware that this is the case and also that, because their activities affect the public interest, they cannot be run in conflict with the wishes of the nation as a whole. In any private industry an unfair profit, one which exceeds the normal returns on capital invested in

Venezuela, is the result of exceptionally skilled management, which harms no one, or in the last resort is only gained at the expense of other private interests. But unfair profits in the oil industry, which handles publicly owned resources, and on which the standard of living of our people and also our hopes of economic, social and cultural development hinge, do unjustified harm to the community when they go beyond a fair return on capital. Private citizens who decide to invest their money in industries of this kind must be ready to accept special conditions, and are never in a position to argue that the high profit rates which might be tolerable in private industries are fair ones under these different conditions. The well-known US writer Arthur Dervin put it brilliantly in an article in *The Annalist*, in January 1928:

> Public service industries have a right to make fair profits, from 6 to 8 per cent on capital invested (if they can reach that level). But they have no right to profits stemming from particularly skilled management of their company. Moreover, if, as a result of incompetence, or simply by accident, the company's administration is not up to normal standards, the losses should be borne directly by the industry itself, and only indirectly by its legal owners. The holders of its bonds and shares have nothing to gain from exceptionally skilled management, and everything to lose if their industry is run badly.

Much the same principles are stated by US government publications, like *Rate Terms and their Meanings* (September 1930), which enunciates the general rule that a public utility company is not allowed to set aside the excess legal profits it makes in good years to make up for losses in previous years.

These principles are now current in public law, because they uphold the public good. But they had already been widely accepted in the field of private law, because of the underlying concerns of that branch of law.

The theory of lesion in canon law, which gradually caught on in positive law, was concerned with this very topic, the definition of a just profit in every kind of human relation. For a time this spread of this theory of lesion was blocked because of a false sense of security, but modern law has rethought this subject completely, and is now having success in making justice more effectively by following highly respected authorities who had already solved the problems presented by some legislation. So, in modern times theories have evolved dealing with the fulfilment of obligation in good faith, and covering unforeseen circumstances, and ones which would make it impossible to carry out a contract. In reality, then, there is no great dichotomy (such as might superficially seem to exist) between the principle of public law based on the common good, and those of private law, based directly on morality and justice.

These are the principles I have used in my campaign to uphold the nation's rights to compensation for the lesion it has suffered at the hands of the oil companies, on issues which are quite independent of the technical details (some more obvious than others) through which many of the concessions could be shot to pieces. Trickery, fraud, bad faith, and lack of foresight and other forms of lesion riddle these contracts; it is also quite clear that they dealt with the exploitation of public assets and have harmed the public interest when allowing profits above normal levels.

We can then argue that this bill is intended to rectify the economic balance between the state and the companies. But if this balance is not in fact a true one for any reason, whether through the negligence or incompetence of those who represent our government, or because of the subtlety or deceitfulness of the private companies, we will find ourselves in an unacceptable and unfair situation, in which the law does not in fact provide any protection. The new concessions granted under this law will, like their predecessors, have dubious validity, insofar as they are against the common good; for public assets, which provide for and guarantee the upkeep of the state, should be exploited and administered in good faith, and the private companies which hold concessions over these assets should be prepared to accept a normal level of fair and equitable profits.

We must now try to discover how far this bill manages to reform the economic bases of the concession system in such a way as to guarantee a just balance which will last through future changes, and to what extent the statistical basis for this formula was in fact taken for granted.

In order to decide what is a just profit for the concession-holder it is necessary to obtain two vital pieces of information: the total value of capital investment and the production cost per barrel. The Hydrocarbons sub-committee of Congress (of which I was a member), in its analysis of the Report submitted by the

Development Ministry for 1941, concluded that 'in order to be able to decide what oil policies to pursue, it is of vital importance for the Executive to work out, as far as it can, what this country should be getting out of the exploitation of its hydrocarbons and other minerals, and also, more important, how much capital the companies have invested and what profits they are making'. But the Development Ministry does not seem to be aware of what this implies. Since then it has done nothing towards finding out what the capital requirements of the industry are, or what it is making, on the basis of properly calculated production costs. The preamble to this bill only explains how the state's share in the profits will be increased; it makes no attempt to do what would be needed to establish a true balance, namely to break down the contributions made by the state on one side and the industry on the other and make their respective shares in the profits dependent on this proportional relationship. It appears that only one Development Ministry official, Dr Edmundo Luongo Cabello, has disclosed the figures which I have quoted above, which were the ones which those who drew up the bill had at the back of their minds. For one statistic, the production cost, no justification is given and there is no estimate of the total capital investments in the industry. When I and some other members of the sub-committee tried to find out how the figure for production costs had been reached, I soon discovered that the Ministry had made no effort to investigate this subject; the officials responsible for this figure had simply accepted an estimate made by some US experts whom they considered to be competent and honest. This figure can be challenged by using the available statistics for capital and equipment brought into this country by the companies, as the records handed over to the sub-committee by the Ministry show. We could not discover what factors the figure produced by the American experts' cost of production estimate includes because no evidence on this subject was submitted to the sub-committee. The available calculations which establish the unit cost only include large figures without any explanation, like, for example, an item of 6 million dollars for profits on sales by the companies in this country, which could be used to cover production costs. Even if we accept all the other figures, the item of 10 million dollars for the salaries of foreign employees, which is very hard to justify, we can reduce the estimate for the total production costs by the 6 million dollars which the companies make in Venezuela by selling their products. That makes a difference of 3 cents per barrel. And if, when analysing this unit cost of 40 cents per barrel, we take into account not only the 3 cents which we Venezuelans pay for the petrol, diesel fuel and other petroleum by-products sold here, but also scrutinize other items, bearing in mind what quantities of capital and equipment have been brought into the country in previous years, it soon becomes clear that those 40 cents include the interest payments on the capital invested, and something to cover the possible exhaustion of reserves.

So it is not quite correct to say that, if a barrel of oil sells for 81 cents, and Venezuela gets 24½ cents, the companies will only make 16½ cents per barrel, because the estimate of 40 cents for the cost of production can quite fairly be said to include 3 cents made on sales in this country and a further amount to allow for depreciation of equipment (which can be sold, because it need not necessarily be handed over when the concession expires) and also for interest payments and reserve funds.

One still more serious criticism of this bill must be made, which reveals what arbitrary statistics the bill is based on, despite its much-praised technical accuracy. Even if we accept the figure of 40 cents per barrel, which, as we have seen, is based on rather vague evidence, the proportions of 24½ cents for the state and 16½ cents for the companies would in fact only work out like that under rather unlikely circumstances, according to present trends.

The bill sets up this hypothetical balance between the state and the companies on the basis of an average selling price of 81 cents per barrel. This was worked out on the basis of the prices for 1937, 1938, and 1939, which are far from typical years since they cover the period of tension leading up to the outbreak of war. The Development Ministry Report gives the figures for these years as respectively 93·7, 82·2 and 76·5 cents, which for a start gives an average of 84 cents not 81. We should also remember that if, as is normally done, we take the 1926 price as an index of 100, the 1936 price works out at only 57·3, which means that the average price is well above 81 (or 84) cents. In 1941 the price of a barrel of oil reached 92 cents, and in 1942–3 it rose to over one dollar; there are good prospects for further rises after the end of the war. Since for every 6 cents above the base price of 81 cents the state only gets one cent more, while the companies get 5 cents under the provisions of this bill, it is quite clear that, with the average price going up fast, the balance which this bill is supposed to set up will in fact soon be upset. The tax on production gives the state only one-sixth of the value of rising prices, and, although the bill allows the Executive to bring down the level of taxation if the profits of the industry after tax go down so much that the industry ceases to be commercially viable, there is no provision for keeping up the proportion which goes to the state when the selling price goes up above the hypothetical level of 81 cents a barrel.

Moreover the fixed taxes (which account for 11 of the state's 24½ cents per barrel) are mainly linked to the surface areas of the concessions. So they do not go up when more oil is produced on the concessions concerned. When production increases, the industry's profits will go up while the state's revenues from this source remain the same.

For these reasons the situation posited by the bill could only occur when an exceptionally low level of production coincides with a very low selling price for oil. To bear this out let us see what would have happened if the bill had been in force in 1941 and 1942. The preamble to the bill gives a table to show what would have happened in 1942; the total government income from oil, including the import duties which would have had to be paid (which is a very tricky item to assess), but not the value of the hectares which the companies are to give up, would have been 113,328,527 bolívares, for 23,554,777 cubic metres of oil produced. The amount from the tax on production is given as 76,638,031 bolívares; if we multiply this figure by 6 (given that the tax on production is fixed at one-sixth of the total value of the oil produced) we get the total value of the oil: 459,828,186 bolívares. If we convert the figures for the volume of oil produced from cubic metres to barrels at 6·28 barrels per cubic metre, we get a figure of 147,923,999 barrels. Multiplying this by the cost of production (accepting the questionable figure of 40 US cents per barrel and changing it into bolívares at the rate of 3·09 to the dollar) we get a total cost of production of 182,834,062 bolívares. If we subtract that from the total value of the oil produced we get a figure of 276,994,124 for the total profit made by the industry; the state takes 113,328,527 bolívares, which leaves 163,665,597 bolívares for the oil companies. That plainly gives the lie to the 24½–16½ relationship which is supposed to be the key factor which sets up a new balance between the industry and the state, and to bring it into line with the most widely accepted principles regulating industries which affect the public interest.

If we make the same projections from the Development Ministry's table for 1941, we get the same paltry results for the government's share in the profits of the industry, once more dead against what this bill is supposed to do.

These are my reasons for voicing my dissent in part from the judgement of the majority of the committee. I believe that if the bill is passed as it stands, and is not modified to iron out the failings to which I have drawn attention, it will have a short life and will need to be revised in the near future.

Caracas, 5 March 1943

Abstention of the Minoría Unificada

The undersigned, deputies of Acción Democrática and of independent political groups, consider it their bounden duty to put together in a single document the objections and criticism they have made in the Chamber during the discussion of the Law of Hydrocarbons.

This is not the time to avoid the issue and shirk the responsibility of analysing the relationship between the state and the oil companies, which is the most important subject which has come before Congress in several years. The whole country expects the members of both Chambers to produce an illuminating discussion of the vital issue of how to regulate the oil industry, which is the hub of our whole economic and social structure.

We are aware that we are under an obligation which affects both the present and the future of Venezuela. So we shall express our points of view with calm objectivity; on this subject as on all others we are convinced that there should exist in Venezuela a political climate in which all opinions which are inspired by the common are tolerated.

In supporting some parts of the bill and attacking others we are doing our duty in two different capacities: first of all as Venezuelan citizens, and secondly as representatives of the people's will, enshrined in our National Congress.

Let us begin by stating that we are all agreed that the bill has many positive advantages in two areas: first, technical and legal reforms; and secondly, changes which will benefit the national economy. In the first area, as deputy Pérez Alfonzo said in his minority verdict on the law, the simplification of the legal relationship between the state and the concession-holders is a great achievement, because it puts an end to the heterogeneous variety of contracts set up under a whole series of changing laws and decrees.

Another legal change of great significance is the decision to give the state tighter control over the industry; in several clauses of this law the state's right to intervene in the different stages of the industry, in the production, transport and refining of oil, is clearly and firmly stated.

The economic gains under the new law include an overall increase of taxes, even though some of the present taxes, like, for example, the tax set up by the law of buoys and lighthouses, will be abolished. Furthermore the exemption from customs duties enjoyed at present by the oil companies is changed from a legal right into a favour which the government can grant only under special circumstances, and the companies are forced to keep their industrial accounts inside this country. The arbitrary distinction for tax purposes between concessions on land and concessions under the sea is abolished under the new law, which includes, last but not least, agreements under which oil refineries will be set up on Venezuelan soil. One of the taxes which this law changes in a way which, in our opinion, benefits this country, is the tax on the surface area of concessions.

We support the controversial Article 40, which relates to this, because it puts up the minimum tax per hectare to 5 bolívares and will also force the companies to give up a considerable amount of the land they at present possess in large quantities because the existing tax on surface area is absurdly low.

Our first main criticism is against the idea that the law 'cleans up all previous defects' in the concessions which are made to fit in with its terms, while the state has to give up all its claims and lawsuits against the concession-holders.

We do not think that this legal fiction can plaster over the cracks which are visible in most of the concessions now being exploited for oil, or compensate the government for all the tax evasion practised by the companies, or give an aura of legality to the enormous profits made by the companies at the expense of a country which has made little progress despite the rapid surge of oil production.

It would be very easy to produce detailed evidence to show what tenous legal bases there are for most concessions, and how the oil companies have systematically broken the laws of Venezuela; for example, all of them have continually avoided paying taxes by fraud or by simple bad faith as in the case of the tax imposed by the law of buoys and lighthouses and several other taxes. Quite apart from these infringements of the laws, the companies should be forced to pay some compensation to the government for the large illicit profits they have been able to make under a legal system riddled with injustices from a Venezuelan point of view, before their present contracts are renewed and extended for another 40 years.

On this subject there can be no argument at all about the figures produced by deputy Pérez Alfonzo in his minority report; during the whole period it has operated in this country the oil industry has made a profit of 3,800 million bolívares, which means that it has been able to recover the capital invested several times over.

It is also quite easy to show that the government could demand large payments in return for the 'purification' of the oil industry. We have only to look back to 1937 to find the Mene Grande Oil Company paying the government (in round figures) 15 million bolívares as a result of an Appeal Court sentence, in respect of taxes it had refused to pay in bad faith; and in 1941 the same company paid into the Treasury 30 million bolívares to clear up doubts surrounding its title deeds to Lot 5 of the López Rodriguez Concession. Going back a little further, we find that in 1932 the Caribbean Petroleum Company paid the government 10 million for the invalid prolongation of the Valladares Concession, and also, as is widely known, gave further large amounts to the lawyers who worked for them and the politicians whose influence had been of assistance.

All this leads us to conclude that the Venezuelan government would be perfectly justified in claiming a sizeable sum of compensation in exchange for the purification of the oil industry, so that it might be able to set about emancipating the Venezuelan economy through an ambitious large-scale programme which would include widening our basic sources of wealth, improving health and increasing population, and providing schools to educate the people.

There are two arguments which are normally used to justify renewing the concessions without insisting on any compensation payments in advance.

The first argument is that the country has already got sufficient compensation by forcing the companies to give up the privileges written into their contracts when these are adapted to the new law. To counter this argument we have only to point out that it is now a generally accepted principle of modern administrative law that contracts which deal with public assets can in no sense be treated in the same way as contracts between individuals concerning privately owned property. From the first type of public contracts 'spring not contractual rights in the normal sense, but merely benefits which are liable to the modifications implied by future legislative reforms' as Dr Nestor Luis Pérez put it when Minister of Development. If it is argued that this is legal doctrine, not legal fact, we can easily give the example of Colombia, which has more than once changed its oil laws; the oil companies there, which are the same ones as those which operate in Venezuela (though sometimes under different names), have had to adapt their contracts to fit in with the new situation produced by the legislative reforms.

Conclusive evidence is provided by the following quotation from the preamble by the Colombian Minister of Mines and Petroleum, Nestor Pineda, to a bill proposed in 1942 for 'State intervention in the oil industry'. This is how Pineda put it:

> Article 8 sets up a tax on the surface area of concessions, to be paid annually in advance by all those who hold contracts to exploit or explore for oil. This

tax will replace the contribution assessed by surface area as set up by Article 19 of Law 37 of 1931 and Paragraph 7 of Article 4 of Law 160 of 1936. It 'shall be paid by all concession-holders, even those whose contracts were signed before the new tax becomes law'. It could also be argued that the tax constitutes a unilateral and therefore invalid alteration of the terms of the existing oil contracts. But it should be remembered that our Supreme Court has already pointed out, with the appropriate legal justification, that new taxes must be paid even by those whose contracts predate the laws which set them up in its decision on a petition for the Colombian Petroleum Company to be declared exempt from taxation under Law 68 of 1935.

The second argument runs as follows: only a victory in the courts would get the government sufficient compensation payments to go with the renewal of the concessions, and it is known from bitter experience that decisions in the national interest are hardly to be expected from the highest courts in Venezuela. However we cannot accept this argument. The Venezuelan people, which is the fount of national sovereignty, has the power to ensure that justice is done in its own interest from the judges' benches.

There are many effective and perfectly legal ways for this country to get what it is fairly due (instead of graciously giving it up). They range from reforming the constitution so as to state explicitly that those who hold concessions relating to public assets are obliged to accept any changes in their contracts implied by legislative changes, to removing judges whose minds are closed to the reformist tendencies in modern law, or whose honesty is suspect.

Having made our position clear on this issue, we will go on to make some criticism of the system of taxation. We will be concerned above all with the new tax on production, since we have already expressed our approval of the new tax on surface area.

The tax on production under the new law is fixed at 16½ per cent of the total value of oil produced, which is the highest rate in the history of our mining legislation. In the laws of 1921 and 1922 the rate was set at 15 per cent, which did not reappear in subsequent laws until 1939, when an oil law laid down a rate of as much as 16 per cent for some types of oil. Let us repeat that this 1943 law contains the highest rate set up by any oil law in this country; it is also important to note that this tax will replace some very low taxes, like the 7½ per cent royalties and the 2 bolívares per metric ton which is in force for concessions which were regulated by malleable legislation or iniquitous contracts.

But, after saying this, we must still ask whether this production tax at a new rate, together with the fixed taxes also laid down by this law, will give the nation a fair share in its oil wealth.

The studies carried out by independent deputies, above all by deputy Pérez Alfonzo, who summarized his conclusions in his statement to justify his abstention, have convinced us that the new tax, though more effective than existing taxes, does not give the nation a fair share in the profits made by the oil companies; nor does it ensure that Venezuela gets more than 50 per cent of the income of those companies.

As Dr Pérez Alfonzo points out, we need two basic statistics in order to be able to work out the true income of the oil companies: the real value of their investments in this country, and the production cost per barrel of oil. The preamble to the bill makes no mention of these vital statistics. When we asked the Development Ministry for technical data we soon came to two conclusions: the officials in the statistical departments of the ministry faithfully accept the estimates made by US experts of 400 million dollars as the total value of the capital investments of the oil companies and 40 cents as the production cost per barrel.

This second figure is very difficult to justify and it should perhaps be pointed out that a law with a wealth of so-called technical achievements like this one is seriously weakened if the statistics on which it is based are not accurate. There is now a very widespread rumour that some of the statistics which the President was given to study are false.

We cannot accept this figure of 40 cents as the production cost of a barrel of oil because it has been worked out on the basis of some absurdly exaggerated figures, like the estimates of 15 million dollars (in round figures) per year for the depreciation of equipment, or 18 million dollars (again in round numbers) for salaries, services and administrative expenses paid in dollars, or 6 million dollars in bolívares earned by the companies from sales of petrol, diesel fuel and petroleum by-products in this country, where in practice they enjoy a monopoly. Even if we accept all the other items, some of which are obviously inflated and some of which can hardly be considered components of the cost of production, such as, for example, payments to executives and directors who live and work outside Venezuela, it is a quite undeniable fact that the item of 6 million dollars for profits on sales in this country has nothing at all to do with the cost of production.

If we set aside this item, the cost price of a barrel of oil goes down 3 cents.

In his statement Pérez Alfonzo also pointed out an even more important fact. He demonstrated that the claim that Venezuela will get at least 50 per cent of the total income derived from oil has no serious statistical foundation at all.

According to data provided by Dr Edmundo Luongo Cábello in one of his lectures, and accepted quite uncritically by those who signed the majority statement of the Chamber's Development Committee, this claim is based on three facts: 1 The average price of a barrel of oil in recent years is 81 cents. 2 The new law sets aside 40 of those 81 cents as the production cost and distributes the remaining 41 cents between the state and the companies. 3 The government will collect 24½ of those 41 cents through taxation, leaving 16½ cents per barrel for the companies.

If this worked it would mean a fair balance, because it would give this country, which is after all the owner of the oil deposits, a larger share than the companies, which only supply capital (which has now been repaid several times over), plus technology and a commercial network in and outside Venezuela.

However, in actual fact the figure of 81 cents is not a proper estimate at all. It is an approximate average of the selling price in the three years from 1937–9, which were untypical years of pre-war tension. The average price in recent years, on the basis of a less arbitrarily chosen period, is well over 81 cents (in fact for the 1937–9 period, according to government statistics published in the Development Ministry Report for 1940, it is 84 cents). In 1941 a barrel of oil sold for 92 cents and in 1942–3 the average price is one dollar. The present forecasts are that the price will stay at this level or may even go up.

Under the new law the further the price of oil rises above the 81-cent estimate, which, we must repeat, is entirely arbitrary, the lower the proportion which goes to the state and the higher the profit rate for the companies.

This is because what the law in fact means is that the state gets only one-sixth of the value of the price rises and the remaining five-sixths goes straight to the companies.

Deputy Pérez Alfonzo worked out what would have happened in 1942 if the law had been in force and was able to show that while the state would have got 133,326,527 bolívares (which is the figure given in one of the statistical tables in the preamble to the bill), the companies would have made 163,665,597 bolívares, not including any profits they might have made on items included in that highly questionable figure of 40 cents for the production cost per barrel.

Because the country will not attain a fair share in its oil wealth under this new law, the rise in oil revenues will not have a great effect on Venezuelan society. In fact, as one of the statistical tables attached to the preamble to the bill says, the increase will only amount to 3·5 million bolívares a month. That is an optimistic estimate, but, even so, would hardly be enough for the government to pay off its deficit, which is now mounting month by month as revenues fall off, not only customs duties and other internal taxes, but even mining taxes themselves, as the volume of oil produced in Venezuela goes down as compared with previous years.

Several other criticisms can be levelled against the law. Above all we should record a protest against the way in which the concessions are extended for another 40 years, after which they can be renewed for a further 40 years. In the USA the maximum length of a concession is normally 20 years, and in the rest of the world the dominant tendency is to reduce the duration of concessions rather than to extend them.

We attempted to change the phrasing of the law, as this Chamber is well aware, by moving some amendments, and by suggesting that some clauses should be studied by special committees. One of the amendments which the majority rejected without giving it serious attention was a proposal to authorize the government to fix the price of petrol and other petroleum by-products on the grounds that they are goods which are vital to the public interest; that would have put an end to the absurdly paradoxical situation in which petrol is sold in Venezuela, the world's third biggest producer and its biggest exporter of oil, at three times the price at which it is sold in, for example, Persia. This amendment was defeated like all the others; some deputies argued that the bill should not be modified because of its peculiar, privileged status as an organic law. But this argument is a very flimsy one because, if Congress was only able to legislate *a posteriori* on decisions which had already been made, it would be giving up the rôle assigned to it by the Constitution, which is the very essence of the republican system.

For all these reasons we believe that there is no basis at all for this emphatic declaration in the preamble to the bill that has now become the Law of Hydrocarbons: 'This bill is not just the basis for another new law; it represents the final consolidation of the economic and technological structure of the industry, along the lines of all those countries which have reached the height of their prosperity.'

This claim that this law 'represents the final consolidation of the economic and technological structure of the industry' cannot be echoed by anyone who believes, as we do, that it is a perfectly legitimate national goal to aim to get the greatest possible advantages out of our oil wealth. We must hope, then, that Congress will in the future exercise national sovereignty on behalf of the Venezuelan people and set to rights those parts of the 1943 Law of Hydrocarbons which are at odds with our country's aims and interests.

Dr Martín Vargas, Andrés Eloy Blanco, Mario García Arocha, Jésus Ortega Bejarano, Víctor Alvarado Franco, Carlos E. Lemoine, Juan Guglielmi, Lorenzo Antonio Vivas, Luis Lander, Germán Orozco Jiménez. Caracas, 12 March 1943 (1)

(*Note* (1): Some years after 1943 when this bill, which basically favoured the interests of the oil companies, was passed, it was conclusively demonstrated that legal advisers of the companies and their contacts in the State Department of the USA played an active part in drawing up this law. In 1976 Dr Juan

Pablo Pérez Alfonzo told the story on the basis of original documents which were found in the government offices at Miraflores when the Junta Revolucionaria took power on 18 October 1945. Pérez Alfonzo, who became Minister of Development (and was therefore in charge of everything which is now classified under Mines and Hydrocarbons) in the provisional government over which I presided, says, when recalling the statement with which he justified his abstention alongside a tiny minority from the vote on this bill in a tightly controlled Chamber which turned the Hydrocarbons Bill into law:

> I did not then know how the nation [He means the government. *R.B.*] had been induced to present the bill in this form. I only found out after the revolution of 1945, when we discovered the relevant documents in Miraflores and learnt that a world-famous firm, owned by Curtis and Hoover, the son of the former president of the USA, had played a decisive part in drawing up the bill. This same firm was later to take part in the agreement in Iran through which, with the approval of the US government, the US companies won the upper hand in that country and the Anglo-Iranian Company was left with a minority holding. It was this firm which was hired by an oil lawyer in Venezuela, as we discovered when we found the manuscript of the studies it carried out in Medina's desk at Miraflores. (General Isaías Medina Angarita, the president who was overthrown on 18 October 1945.) (See J. P. Pérez Alfonzo, *El Desastre*, Vadell Hermanos, Valencia, 1976, p. 51.)

In 1954, after the overthrow of that great man Mosadegh, this group carried out a piece of financial wizardry in Iran, in which the very close interrelations between oil and diplomacy came out in a very striking way. For his active rôle in that rather peculiar agreement Herbert Hoover Junior got a very significant regard from the oil-conscious Eisenhower administration. He was in fact made Under-Secretary of State to Foster Dulles to help form the USA's foreign policy at the highest level.

Let us go back to the vaunted Venezuelan Hydrocarbons Law of 1943. While these foreigners were drawing up the law behind the scenes, a public commission of legal experts was set up and they were consulted about each article, but without knowing who was producing it. As Pérez Alfonzo says: 'I have talked to several members of this commission, and they have all told me that they had no idea that a US firm was drawing up the bill; they assumed that the Ministry was drawing up the articles and then submitting them to the commission for analysis or revision.' (Ibid. p. 52.)

Max Thornburg, the Oil Adviser to the State Department, also took a hand in drawing up this bill. As the former Secretary of State Cordell Hull admits explicitly in his memoirs, Thornburg was working for the oil companies, who had got him appointed to protect their interests. Hull says that he appointed Max Thornburg as Oil Adviser on the recommendation of Dr Feis, the director of the State Department's International Oil Policy Committee, and later adds: 'When we learned that Thornburg was still connected with a US oil company, I immediately requested resignation.' (*The Memoirs of Cordell Hull*, Macmillan, New York, vol. II p. 1,517 quoted in R. Betancourt: *Venezuela, Política y Petróleo*, Editorial Senderos, Caracas 1967, p. 194.)

Declaration by the Confederación de Trabajadores de Venezuela

The CTV and the nationalization of oil

The promulgation of the Law which reserves for the State the production and marketing of hydrocarbons marks the end of a long period of Venezuelan history. Fifty years of exploitation have come to an end. Today Venezuela is taking control of its own oil industry, and, with it, the enormous responsibility of running it for the benefit of the people.

On this historical occasion, the Confederación de Trabajadores de Venezuela (Confederation of Venezuelan Workers), which has always fought for the nationalization of oil as a vital step in the economic and social development of our country, feels great satisfaction and a legitimate sense of pride.

There has been a thorough debate in Congress about the implications of this law. The different parties have all taken the opportunity to express their points of view. When Congress has sanctioned the law which the President has now signed, and the sovereign will of the Venezuelan people has thus been upheld, all the patriotic feelings in this country should be channelled towards supporting the nationalized industry, and defending and watching over it to ensure that it is run in accordance with the basic needs of our people.

The workers for whom the CTV speaks see the

nationalization of oil as the means to emancipate the national economy; it will bring sweeping changes to make our oil resources help Venezuelan economic development, and, still more important, to carry out a more even redistribution of national wealth and put an end to the shocking social inequalities produced by the distortions involved in having our oil wealth in the hands of foreign economic interests.

We hope that the state companies which will be set up to run the oil industry will carry out decisive policies with a definite social content, so that the men and women of this country can fulfil their true potential. Now that this oil wealth is under the state's control, it must no longer be enjoyed by the minority which has always squandered non-renewable natural resources like oil for its own short-term benefit.

There must be no repetition in the nationalized story of what happened in the Corporación Venezolana de Guayana: this company had great success in terms of productivity and the volume of its production but it can hardly be denied that it failed to take account of the basic social and economic factors of man's existence, and did nothing about the standards of living or the labour conditions of its workers. These basic factors should be the main concern of every government, which aims, as that of President Carlos Andrés Pérez does, to bring social and economic democracy into existence inside the framework of liberty and full human rights.

The Confederación de Trabajadores de Venezuela and the workers who belong to its member unions do not hesitate to express their support for the nationalization of oil. We are ready to defend the industry from every one of the difficulties and dangers which will no doubt threaten it as a result of the firm decision we have taken as a nation and as a people in connexion with a vitally important commodity in the world economy. We hope that our nationalized oil will make a contribution to the progress and development of the Third World countries, which have high hopes of Venezuela.

The labour movement will keep a responsibly vigilant attitude towards the nationalized oil industry, and towards those who will direct and administer it in the name of the state; our aim will be to give concrete form to the principles of social and economic democracy, for which we are constantly struggling.

The Confederación de Trabajadores de Venezuela is well aware of the economic realities which will affect the nationalization of oil. There will be difficulties during the formation of a nationalized company, but we are confident that the skill and patriotism of our technicians and workers will help us to overcome the obstacles ahead without creating unnecessary problems; they will make their claims and complaints known through the legitimate procedures of our organization, so that solutions can be reached after the fullest possible discussions between the nationalized company and the workers.

We are patriotic workers, who are dedicated to our task which is vital to the present and the future of Venezuela, and we want to make an active and effective contribution to these events, not a purely formal, decorative one.

The nationalization of oil presents a great challenge to our country. But the men and women of Venezuela tend to grow in stature when confronted with difficulties. So, whatever extremes they may reach, the half-hidden political intrigues surrounding the process of nationalization will have no effect.

The nation has taken a far-reaching decision. As the vanguard of our people, the Confederación de Trabajadores de Venezuela appeals to all Venezuelans for unity so that we can set out with optimism on the historic journey towards our economic emancipation, on which we are making a decisive start today.

José A. Vargas, President; Rafael León León, General Secretary; Casto Gil Rivers, Administrative Secretary; César Gil, Labour Secretary; Andres Hernández Vázquez, Financial Secretary; Manuel Peñalver, Secretary for International Relations; Carlos Luns, Secretary for Press and Publicity; Francisco Urquis Lugo, Secretary for Employment and Professional Training; Rubén Santiago, Secretary for Technical and Contractual Studies; José Beltrán Vallejo, Secretary for Labour Representation; Lamario González U., Secretary for Union Education; José González Navarro, Secretary for Parliamentary Affairs and Social Legislation; Máximo Acuña, Secretary for Economic and Social Activities; Armando González, Secretary for Agriculture; Antonio Ríos, Secretary for Culture and Recreation.

Committee Members: *Isaac Olivers, Angel Zerpa Mirabal, Ramón Petit, Alejandro Freites, Andrés Agelvis Prato, Jesus Urbieta, Rafael Domingo Campos, Augusto Malave Villalba, Dagoberto González, Juan Delpino, Pedro Brito, Federico Ramírez León, Juan Díaz, Rafael Castañeda.*

The President's Seal of Approval

The Nationalization of the Oil Industry
(Speech by President Carlos Andres Pérez, in the Oval Room of the Palacio Federal Legislativo.)

On 6 December 1974 I announced to the nation from this Sanctuary of the Fatherland a decree by which the exploitation of iron ore in this country was reserved for the State, and the existing concessions were squashed as from 1 January 1975. Then as now this chest was open and the original declaration of our Independence was visible, so as to show that we are carrying forward the principles of liberation established by the founding fathers of our Republic.

Today, which will surely come to be commemorated as one of the great national days in the year, I have put my seal on the Law which reserves for the State the production and marketing of hydrocarbons. This decision has been ratified by a consensus of opinion which has its roots in a long period of nationalist initiative during which our national consciousness has blossomed. This is the culminating point of an era during which Venezuela has found a new future.

The whole country is participating in this event. As head of state, I am merely interpreting and executing a decision which has been made by the whole people of Venezuela. The rest of Latin America, and all those countries which are victims of the economic totalitarianism of the great industrialized countries will back us up for taking this daring step with such a calm sense of responsibility, because it is right in line with the historical trend of great struggles by the Third World countries.

A promise and its meaning

On 12 March 1974, when I swore my presidential oath in front of Congress, and took up the office which the people had conferred upon me, I made a speech which included the following promise, which I can now say that I have kept:

> The seventies will be a decade of great achievements for Venezuela and for the whole of Latin America. Here, in Congress, in December 1970, our campaign for control over the oil industry burst into flames when we recovered control of the selling price of our basic wealth. Now Venezuela has the opportunity, with the backing of its oil industry, to offer practical assistance to the Latin American countries in their struggle to achieve independent development, higher prices for their raw material exports, and a fair and balanced position in world trade. Today nationalism means no longer simply adventurous rhetoric, but a coherent method of analysing and carrying out policies on behalf of this country and its people, inside the framework of the interests of Latin America as a whole.
>
> We are going to fulfil the old longing of our people for our oil to be Venezuelan. There are several legal instruments in existence which would have guaranteed that the concessions return to the state. But

we have judged this to be the right moment to speed up the process by deciding on a new national and nationalistic oil policy, which must be debated and finally approved in front of the nation as a whole. It is Congress which will have the last word, on behalf of the whole nation, not merely on behalf of a parliamentary majority.

What we need is not so much a new law as the agreement of all of Venezuela about what we can and should do to carry out the unique task allotted to us by history. My government's oil policies will never be distorted by dogmatism of any kind. I believe that the nationalization of oil is not a rhetorical device but a plan of action. The softer our protests the easier it will be to make ourselves heard and understood. If we begin to shout we run the risk of setting up conflicting echoes which may confuse us and lead us astray.

I have set up a wide-ranging commission, composed of important national figures, to advise the government as we study the plans which will later be submitted to Congress. In this way I hope to be able to win the active support of all Venezuelans and form the widest possible consensus for the all-important decisions we will have to take.

Ten days after this speech, on 22 March 1974, I signed the decree which set up the Presidential Commission to help the government study the measures to be taken in order to gain control of oil exploration and the exploitation, manufacture, refining, transport and marketing of hydrocarbons. When the members of the commission took their oaths I said:

We are undertaking a task which has enormous implications for the present and future of Venezuela. It requires the united support of all Venezuelans. These are not just government decisions, or simple practical and legal problems which have to be solved before the industry can change hands. Later on we will be taking responsibility for the efficient management of an industry of far greater dimensions than anything we Venezuelans have had to grapple with before with our clumsy and cumbersome bureaucracy and our lamentable lack of men of great ability and a strong spirit of public service who will be needed to take the vital decisions which will dictate the future of our oil industry. We will take every precaution not to compromise our future by taking hurried decisions or taking up disorderly demagogic positions. It will be necessary to look for and make agreements all round so that everyone can feel part of our campaign. We must look also at what has happened in the OPEC countries and in the European and Latin American countries where state companies have been set up and maintained to operate the oil industry, and seek their advice. For Venezuela must learn from all of them. In this world lessons are learnt by trial and error.

In charge of our own resources

On 23 December 1974, the report of the Presidential Commission was presented to me, and, when I thanked its members for their services to our country, I said: 'I am absolutely certain that, when we have studied this report, we will propose to Congress a bill which follows very closely, if not exactly, the recommendations contained in this report.'

All these promises have been kept. In the ceremony which I have just performed in front of the highest representatives of the powers of state, and of all sectors of Venezuelan society, I have given my authority to the legal instrument which puts the vital decisions in our hands, so that we can in fact take full possession of the production and marketing of hydrocarbons on 1 January 1976, just as we took control over iron ore on 1 January 1975.

The Venezuelan people has decided to make history by giving reality to its unanimous desire for sovereign control over its natural wealth. On this occasion above all others, I am representing the whole country, including both those who have given full support to our bill and those who have expressed disagreement with some aspects of our initiative. There can be no division which sets us apart in this common responsibility shared by all Venezuelans.

Over these last months I have followed with great interest and enthusiasm the passionate debates between different political groups in Congress occasioned by the Executive's bill, and also the comments and the disputes it has caused in the nation as a whole, as recorded by the mass media. We can point out with a patriotic sense of satisfaction that the discussions inside the Presidential Commission, which covered the widest possible variety of ideologies in this country, and the debates in Congress (though they were lengthy and contradictory) all helped to form the indispensable consensus which I had requested on the day on which I took office, and which today is borne out by the unqualified support of the whole people.

An opportunity which had to be taken

There is no question for any Venezuelan that this is the right moment, the opportunity which had to be taken, for us to take full control of the production and marketing of hydrocarbons. The inevitable quite salutary disagreements have sprung from different ideas about the way to reach the overall goal of nationalization. The law which we have just promulgated contains no statements or ideas which contradict or might be turned against our nationalist principles or the underlying interests of this country and its people. I have great respect for dissenting opinions, and understand the doubts and even fears which some political groups may feel about the way in which the executive over which I preside will put into practice the changes which the law entails. Even though, it should be remembered, any decisions which might involve dangerous changes in the nationalization plan will have to be approved by both houses of Congress in a special joint session, I accept full responsibility, as it is my duty to do, for proving by my future actions that I do not intend to budge an inch from the vital aims which Venezuela has sought and has achieved now that iron ore and oil have been nationalized.

The disagreements about the details of the law arose freely and openly, as I wanted and had suggested they should. But, in a way that has nothing to do with those passing controversies, my compatriots are fully conscious of the fact that, on questions which affect our national sovereignty, the country needs to be united and totally, alertly aware of its responsibilities.

I am quite certain that my government will be ready to dedicate all its decisive and creative energy on the great struggle to conquer our economic liberation on which we are embarking today. So I am not hesitant or afraid to take up this responsibility on behalf of our country. The success of our plans will undoubtedly take us towards the formation of a greater Venezuela in which we are complete masters of our own destiny. It is my duty, as a democratic ruler, to make use of criticism from the opposition as a stimulus, or as a warning signal which shows that the country is keeping an alert watch over every move my government makes on the path towards a new Venezuela. Today's ceremony represents an act of sovereignty, which should be remembered as the start of a period of constructive creativity.

This will be a test of our ability and maturity as a nation; no longer will we get easy benefits from the profits made on oil belonging to outsiders. We will have to make a great effort on our own so as to get a higher standard of living and more social welfare for all Venezuelans.

Our wealth in our own hands

We shall undertake this great task with daring, and without any extraneous factors to complicate the issue. We have taken our decision without being influenced by political dogmas or by the strategic interests of continental or world powers. We have found our own way to carry out this nationalization without any rash proclamation, as befits a mature rational country which has no intention of allowing its oil to create situations of dependence or to provoke international conflicts.

Now that we are fully in control of our iron ore and our oil, this government must take up its pressing responsibilities in the struggle for the economic liberation of Latin America. We will do so bearing in mind the principles of representative democracy laid down in the Constitution sanctioned by the people in 1961. History will record the story of how this developing country, a constitutional democracy, took control in a truly constructive way of its basic industries without aggressively snatching back what it was due. Our method of carrying out our democratic nationalist revolution has not been copied from anyone. We have produced our own legal instrument to do this without making any concessions which would curtail our liberty or diminish our national sovereignty.

We are aware that all the decisions we take from now on will affect the whole country.

I know that I have the full support of my people, despite the differences of opinion over the law we are putting into force today. So I will not let that consensus of opinion down; anyone who thought that this law might leave loopholes which might compromise our sovereignty or force us to bow down to the foreign interests which have been exploiting our natural resources can rest assured of that. Our firm, upright stand will affect not only our people, but also all the peoples of Latin America and the Third World, that wide-ranging group of countries linked by increasing ties of which we form part.

Venezuelan nationalism has expressed itself openly and without aggressiveness, and has sought international cooperation, understanding and friendship between all countries. We are setting the example for a new kind of international solidarity based on the rejection of all the humiliation and exploitation which our peoples have suffered.

Using oil to open a dialogue

Oil is now a world political and economic problem which involves Venezuela in an ever more complex foreign policy. Oil is the weapon used by the Third World countries which belong to OPEC to force the industrialized countries into a new kind of dialogue which might make it possible to set up a new economic world order. Venezuela is totally in agreement with this campaign for international justice, quite apart from being bound by her close ties with her sister countries in Latin America.

The destiny of this country is changing and we are becoming aware of the great future which could await her. We have shaken off the bonds of conformism, and have got to the bottom of our problems; we are now 'learning to live with the risks' which fully independent sovereignty entails. All Venezuelans, both in the public sector and in the private sector, must now realize the serious dangers we face as a result of this decision. Only if we are aware of this will we be able to maintain sufficient spiritual drive and strength of purpose to face the complicated problems which await us.

First of all we must make sure that optimism prevails over pessimism. In the past sociologists have concluded that we are doomed to be governed by dictators on the grounds that we lack suitable material for a democracy. Today there are similar suggestions that we are incapable, as a nation, of managing our own oil industry.

The Venezuelan people had shown that the predictions made by sociologists and pessimists are false. All through our history we have pursued ambitious ventures. In recent years, in fact since we have had a democratic government, Venezuelan youth has turned towards science and technology to such an extent that we can now confidently say that Venezuelan technical experts and managers we bring in, will ensure efficiency and continuity in the running of the nationalized industry. Of course we do not mean to underestimate the contribution which foreign technical experts will still make. Still less do we mean that we fail to recognize our situation of technological dependency; we shall have to overcome this situation just as we have overcome previous obstacles to our independent development.

A great transformation

The decisions we have taken will require all our energies for many years to come. It is my government's privilege to begin this great radical transformation of our country. All I can do is to set this process in motion; I shall take great care not to let anything for which I am responsible limit any of our country's great ambitions. What must be strengthened is our people's faith in its ability to take up the responsibilities which, until very recently, it was quite ready to leave in the hands of foreigners from the countries which exploited our wealth.

At this historic moment we have come face to face with our country's future; success or failure depends not on the government alone but on the country as a whole. We have given up living on the easy money which the oil bonanza brought us. This is what must bring all of us together; it cannot be the subject of disputes between parties or conflicts between different sectors of our society. This task of building the nation's future is a common objective in which all of us must take part. Now, 165 years after the foundation of this Republic, we are still learning the same painful lesson. Bitter party rivalries and personality conflicts have caused many frustrations for our people.

Our hopes lie in the self-negation and altruism which befits a republic, and the honest courageous endeavour which will forge our destiny. Above all we need foresight, which will enable us to analyse and decide on our nation's priorities, to help us to administer our natural resources. But it is a difficult and dangerous task. To calm our fears we must be absolutely certain that this great effort will help us to reap the benefits of these great national decisions.

The government is not concerned only with economic problems. It understands that there are other equally important priorities which are vital if Venezuela is to become a truly sovereign nation.

This morning I put my signature to the National Law of Culture, thus fulfilling a promise I made during my election campaign, which I wanted to carry out at the same moment as our epoch-making oil nationalization. There are many historical examples which show that the peoples which achieve their ambitions and bring out their national characteristics to the full are those which have left a permanent mark on world culture. It is always the intellectually alive country which is best able to determine its own future. Literature, painting, music, theatre and films, and all the other means by which the human spirit expresses itself, form a vital part of the infrastructure of authentic nationalism.

A democratic revolution
My government has given concrete evidence of its concern for culture as one of the bases of the democratic revolution which it is trying to carry forward. Today I want to bring artists and intellectuals into the public celebrations over the nationalization of oil by giving them the Law of Culture, and allotting 5 per cent of the Gran Mariscal de Ayacucho scholarships to the study of arts and the humanities; this will not reduce the number of technical experts we produce but it will contribute to our national goal of building a nation, the spiritual values of which are dominated by democratic humanism, at the service of the men and women of Venezuela.

It is because we have successfully built up a democracy which is capable of taking these far-reaching decisions that Venezuela is now able to face this great challenge with confidence. We have attained sufficient maturity to be sure of our own abilities. This should be hailed as a collective decision, not the victory of any political group or party.

Enough has already been said about the history of our oil, about the continuous struggle which is coming to a climax today, about the contributions made by many Venezuelans of all political hues to our victory in this campaign. Every clause of the Law which reserves for the State the production and marketing of hydrocarbons has already been subjected to detailed, sometimes passionate criticism. All of Venezuela has been made aware of the difficulties and dangers we face. The opposition parties have drawn attention to their suspicions and fears. The whole country has followed the great debate of our time with baited breath. Now comes the time for action to be taken. I am quite certain that those who have expressed doubts and fears will stop blaming my government when it demonstrates in practice, with a clear and honest attitude, that the facts will not force us to bend our policies in this first stage of the campaign to give Venezuela an independent future. All my countrymen must be well aware that, if we are successful, it will be a triumph for the country as a whole, and the crowning victory for our democratic system and the political parties which uphold it. But it should also be borne in mind that my government will be held responsible if our efforts flag and the law we are putting into force today soon meets with failure. I am perfectly aware of that possiblity and accept full responsibility. All I ask is your confidence and support once you have passed judgment on the decision I take as head of state.

Producing our own wealth

All Venezuelans are equally involved in this process through which we will come to produce our own wealth. This can only be done if the national consciousness which I am demanding is maintained and strengthened. It will need hard work and many sacrifices. By taking our oil industry out of foreign hands we are in fact making ourselves still more dependent on oil. If we do not take advantage of all our natural resources to set up a solid economic structure, we will be failing to realize the implications of this historic event.

We have acquired heavy new responsibilities. The whole of Venezuela, including its political parties, its businessmen and workers, even the foreigners who live here and contribute to our national development, will have to work hard to bring permanent sources of wealth into existence. This is the best opportunity for Venezuela to confirm her existence as a free country and to move towards her destined greatness.

We have borne all this in mind while planning with great care to transfer the administration of the oil industry into government hands. Before the ink of my signature on this law is dry I shall set up by decree a company called Petróleos de Venezuela which will become responsible for carrying out the government's policies and for planning, coordinating, supervising and controlling the companies which will actually run the national oil industry. As soon as we are given the names of the Commission delegated by National Congress, we will set up the Commission to supervise the Production and Marketing of Hydrocarbons, which will control all the activities of the multinational companies between now and 31 December. In the next 45 days we will try to agree on detailed terms for the compensation to be paid, under the law, to the companies which hold concessions. The agreements we reach will then be submitted to Congress for its approval. If we do not reach agreement, we shall have no hesitation in asking the Supreme Court to determine how much compensation we should pay.

Let me repeat, on this historic occasion, that Petróleos de Venezuela will be unaffected by any changes in national politics. It will serve the country's overall interests without taking into account any temporary political influence or personal interests. Tomorrow I will announce the names of the upright and worthy citizens of this country who will head this company.

On 5 July, from this rostrum, in my message to the Venezuelan people, I gave a detailed account of how the national oil company will work; those details are now enshrined in the legal instruments, which Congress and the council of ministers have approved.

A firm policy

This government's policy of conserving our renewable and non-renewable natural resources will be continued firmly and actively in connexion with hydrocarbons. Let me repeat what I said when I came into office, that I will block all plans to export liquid gas. We must keep our gas for our own industrialization, as a great reserve for the future of the Venezuelan petrochemicals industry. In our eighteen months in office we have reduced the amount of gas wasted when refining oil significantly, so that we are now using 98 per cent of the gas produced in this process.

Venezuela can rest assured that everything is ready for this new era in our history. All that remains for us to do is to realize our ambition to make the oil industry into a success which will demonstrate our efficiency and our ability in a way which helps to accelerate and secure our country's independent economic development; for this we will need hard work and vigilance from everyone.

Men and women of Venezuela, we must not, in the euphoria of this great moment, forget that we are only just setting out on the road towards our economic independence. It will be a hard grind before we get anywhere but now we will have no excuses to make for our failures. This is a job for us alone, and the wealth we can produce will be ours. But now we will also become more to blame for destitute children, our poverty-stricken homes, and our abandoned fields.

National consciousness

Our national effort requires the formation of a working class which is conscious of its rights, as a consistent and active social force which will be able to ensure that the profits of our national wealth are not concentrated in the hands of a few, or squandered by the state. It should also be able to ensure that there are no specially privileged workers. We must now override particular interests or interest groups and lay the foundations of a society which will provide true justice for all Venezuelans.

The workers' social welfare contributions, as laid down in labour legislation and in the collective contracts, should be paid into the Banco Central within fifteen days from now, and should be worked out on the basis of the salary of each worker on the day the concessions come to an end. However, the trust funds set up under the recently modified Labour Law in accordance with what was agreed between the concession-holders and the workers for the moment at which this law is put into force will continue to exist. In this context it should be pointed out, as the oil nationalization law says, that the change of management will not produce a change in the relationship between the companies and their employees.

Houses for the workers
It is not in this country's interest, nor in the interest of the state companies which are to run the nationalized industry, nor in the interests of the workers themselves, that this government should go ahead and hand over the houses rented by some of the workers to their occupiers in full possession. We will have first to study the whole problem and then produce a housing plan which will cover all the workers in the industry and put it into effect. If we do not do this, it will simply be flagrantly unjust to other equally deserving workers and the state companies will be left with a serious problem on their hands.

The prices of raw materials
The economic system which made it possible for the natural resources of the poor countries to be exploited will soon be at an end. The Third World countries are no longer prepared to accept dirt-cheap prices for their raw material exports. A new world economic system is in the process of formation. Meanwhile, the industrialized countries don't seem to realize, or to be able to accept, that the countries they have been exploiting can take their own decisions and vigorously defend their interests.

I want to send a message to my fellow Latin Americans as well as to the people of Venezuela, and to proclaim with optimism and great faith in the future that Venezuela's destiny lies in the direction indicated by these words of our Libertador Bolívar, spoken at the Congress of Angostura: 'Only democracy can bring absolute liberty.' Bolívar went on: 'But what democratic government has been able to achieve at the same time power, prosperity, and permanence?' In this country we shall be able to show that his doubts were unfounded by setting an example just as he did. Our aim is to help powerful, prosperous and permanent democracies for all the people of Latin America.

It's our oil

It's our oil and we now have the chance to show that we are competent to run the industry. If we are confident in our ability, oil will help to bring democratic development together with social justice.

Venezuelan oil will contribute to Latin American integration, world security, human progress, international justice and a balanced mutually-dependent economic system. It will also become a symbol of Venezuela's independence and her national will, and of the constructive dynamism of her people. Finally it will take us a long way towards the fulfilment of our destiny. There is no more fitting place to say this than in the presence of Simon Bolivar who taught us to believe in our people, and fought to show what we are capable of.

As we pass an important milestone on the path towards our destiny, let us repeat once more that our unshakeable object is to show that a just society can only come into being if respect is paid to human liberty. We are irrevocably committed to this great campaign through which we are taking the first steps towards the economic liberation of Venezuela; I can only exhort all my compatriots to strive towards this common goal. Let's get to work!

Caracas, 29 August 1975

The Organic Law which reserves for the State the production and marketing of hydrocarbons

The Congress of the Republic of Venezuela hereby decrees the following:

Organic Law which reserves for the State the production and marketing of hydrocarbons.

Article 1: For reasons of national convenience, all that which relates to the exploration of the national territory in prospecting for petroleum, asphalt, or other hydrocarbons, shall be reserved for the State; as well as the exploitation of deposits of the same; their manufacture or refining, and transportation by special means or storage; the domestic and foreign commerce of the exploited and refined substances, and whatever works are required for the handling thereof, as provided herein. The concessions granted by the State shall therefore be terminated. Such expiry shall take effect as of 31 December, nineteen hundred and seventy-five.

The activities referred to in this article, as well as any work or services required to discharge the same, are hereby declared to be of public utility and social interest.

All matters related to the natural gas industry and to the domestic marketing of hydrocarbon derivatives, shall be governed by the Law Reserving for the State the Industry of Natural Gas and by the Law Reserving for the State the Exploitation of the Internal Market for Hydrocarbon Derivatives, respectively, in so far as they do not conflict with the provisions of this Law.

Article 2: The foreign commerce of hydrocarbons shall be under the exclusive management and control of the State, and shall be exercised either directly by the National Executive, or through existing State agencies or any such as may be established in order to carry out the purposes of this Law.

Article 3: The management of the foreign commerce of hydrocarbons shall have the following basic objectives:

To attain the highest possible economic yield from petroleum exports, in accordance with the requirements of national development; to secure and maintain stable, diversified and adequate foreign markets; to support and promote the development of new export lines in Venezuelan products; to ensure suitable terms for the supply of materials, equipment and other elements of production as well as the basic commodities the country may require.

Article 4: In negotiations for the sale of hydrocarbons in foreign markets, the National Executive or the state agencies may, while reserving their right of commercialization, use a variety of methods, designed above all to secure and maintain direct markets for Venezuelan hydrocarbons.

Article 5: The State shall exercise the activities set forth in Article 1 of this Law directly through the National Executive or by means of agencies of its own. It may also sign any operative agreements that may be necessary for a sound performance of its functions.

By no means, however, shall such negotiations affect the substantial nature of the activities hereby attributed.

In special cases, and whenever it suits the public interest, the National Executive or the said agencies may, in exercising any of the activities herein referred to, enter into partnership agreements with private entities; maintaining, however, such a share therein that will guarantee control by the State, and establishing a fixed duration for such agreements.

Authorization from both Chambers of Congress meeting in joint session shall be required prior to signing any such agreements, under such terms as may be set therefor, and once the Chambers have been duly informed by the National Executive with respect to all matters pertaining thereto.

Article 6: For the purposes referred to in the foregoing Article, the National Executive shall organize the managements and administration of the activities hereby reserved pursuant to the following Bases:

(i) Within such juridical forms as it may deem suitable, it shall establish any enterprises it may consider necessary for the regular and efficient performance of such activities and it may assign to them the exercise of one or more such activities; modify their objectives, merge or associate them, abolish or dissolve them, and convey their capital to another or several other such enterprises. Such companies shall be owned by the State, without prejudice to the provisions of Basis No. (ii) to this article, in the event that the same should be organized as joint stock companies, they may be incorporated with one sole partner.

(ii) It shall assign to one of the enterprises the

functions of coordinating, supervising and controlling the activities of the others, and it may ascribe to the same the ownership of the shares of stock of any such other companies.

(iii) It shall carry out the conversion of the Corporación Venezolana del Petróleo, established pursuant to Decree 60 of 19 April 1960, into a mercantile company.

(iv) It shall incorporate, or cause to be incorporated, for the sole purpose of expediting and facilitating the nationalization process in the petroleum industry, any enterprises it may deem suitable, which, upon expiry of the concessions, shall become the property of the enterprise provided for in Basis (ii) of this article.

(v) In order to provide the enterprise referred to in Basis (ii) hereof with adequate resources for the development of the national petroleum industry, the operating companies incorporated pursuant to Bases (i), (ii) and (iv) hereof, as the case may be, shall deliver to the former every month an amount of money equivalent to 10 per cent of their net receipts originating from petroleum exported by them during the immediately preceding month. The amounts thus delivered shall be free from payment of federal taxes and assessments and will be deductible by the operating companies for income tax purposes.

Article 7: The companies referred to in the foregoing article shall be governed by the provisions of this law and the regulations thereof; by their own bylaws; by the resolutions issued by the National Executive, as well as by any applicable provisions of the Common Law. Furthermore, they shall be subject to the payment of the federal taxes and assessments established for the hydrocarbon concessions; to any other applicable rules that may be included with respect to them in the laws, regulations, decrees, resolutions, ordinances or official circular letters, as well as to the stipulations contained in the agreements made between the concession-holders and the National Executive. They shall not be subject to any kind of tax levied by state governments or by the municipalities.

Article 8: The directors, managers, employees and workers of the companies referred to in Article 6 hereof, including those of the Corporación Venezolana del Petróleo once the latter is converted into a mercantile company, shall not be considered public officials or employees.

Special Paragraph: Without prejudice to the stipulations of this article, the provisions of Articles 123 and 124 of the Constitution shall be applied to the directors or managers herein referred to.

Article 9: A Supervisory Committee of the Industry and Commerce of Hydrocarbons is hereby established, responsible to the Ministry of Mines and Hydrocarbons. It shall consist of 9 members, 2 of them shall be designated by the President of the Republic from a list of 3 candidates presented by Congress, or in its absence by the Interim Commission thereof, and 7 of which shall be appointed directly by the National Executive, all of this to be carried out within a period of 10 days counted from the date of enactment of this law.

Two of the 7 members named directly by the National Executive shall be selected from a list of 5 presented by the federation of labour unions which represents the majority of the workers. The object of the Supervisory Commission shall be to exercise the representation of the State in all matters involving the activities of the concession-holders, for the purposes of inspection, control and authorization, until such time as the state enterprises provided for in this law shall assume the exercise of the industry and the Supervisory Commission shall be constituted not less than 5 days following the expiration of the period set forth in the first part of this article. It shall hold its meetings validly when at least 7 of its members are present, and shall adopt its decisions by a majority vote of the members present.

Article 10: By means of a resolution to be published in the Official Gazette, the Minister of Mines and Hydrocarbons shall determine within 10 days of the enactment of this law the matters which shall be subject to inspection and control by the Supervisory Commission of the Industry and Commerce of Hydrocarbons, as well as which acts and decisions of the concession-holders shall require prior authorization by the aforementioned Commission.

Inspection and control shall be primarily exercised over all operating, financial and commercial planning and practices of the companies; over the labour systems and practices of the same, as well as on the costs of the petroleum industry. The functions involving authorizations shall be primarily exercised in connexion with contracts with sales and interchange of crude oil and derivatives, with the remittance of funds and payments abroad, with investment budgets, and with contracts relating to the transfer of technology. This enumeration shall not restrict the powers which the Executive holds by virtue of the existing laws or those powers which may be determined by the Ministry of Mines and Hydrocarbons in compliance with the provisions of this law.

Article 11: For the purpose of the proper performance of their functions, the Supervisory Commission of the Commerce and Industry of Hydrocarbons, or any of its members duly authorized by the Commission, or the auxiliary officials who at the request of the Commission may be appointed by the Minister of Mines and Hydrocarbons, shall have free and unrestricted access to all the works and offices of the concession-holders, to the managing and administrative organizations of the same, as well as to their accounting books and records.

The concession-holders shall render to the aforesaid Commission, and to the members or the auxiliary officials thereof, every facility and cooperation for the sound discharge and exercise of their functions.

Article 12: Within 45 consecutive days following the date of enactment of this law, the National Executive, through the agency of the Minister of Mines and Hydrocarbons, shall make a formal offer to the concession-holders with regard to compensation for all of the rights which they hold over the property subject to their concessions. Such compensation shall be calculated pursuant to Article 15 of this law and paid in accordance with the provisions of Articles 16 and 17 hereof. The concession-holder shall reply to the offer within 15 consecutive days following the receipt of the communication from the Executive. The agreement, if any, shall be spread upon a Memorandum of Agreement, signed by the Attorney-General of the

Republic in accordance with the instructions which for such purpose are given him by the National Executive through the agency of the Minister of Mines and Hydrocarbons, and signed by the respective concession-holder; to take effect as of the date of expiry of the concessions as provided in Article 1 of this law. The National Executive, through the Ministry of Mines and Hydrocarbons, shall immediately submit the aforesaid Memorandum for consideration and approval to the Chambers of Congress meeting in joint session, and the latter shall decide thereupon as early as possible and not later than thirty consecutive days reckoned from the date of receipt thereof.

The Memorandum of Agreement herein provided for shall serve the State as a title deed to the rights and assets object of the Agreements.

Special Paragraph: The persons who shall have entered into agreements for the joint operation of concessions or for participation with companies holding hydrocarbons concessions, shall be subject to all the provisions of this law, and for the legal purposes thereof will be considered as having the same rights and obligations inherent to the concession-holders.

Article 13: In the event that no agreement has been reached under the foregoing article, within 30 consecutive days following the date on which the concession-holder has given notice of his decision not to come to an agreement, or following the termination of the time allowed therefor if the concession-holder has failed to reply, the National Executive shall through the agency of the Minister of Mines and Hydrocarbons, instruct the Attorney-General of the Republic to bring suit within the following 30 consecutive days before the Political and Administrative Division of the Supreme Court of Justice for the expropriation of all rights enjoyed by the concession-holder over all properties attached to the concessions to which it may hold title, pursuant to the following special procedure:

a. The expropriation petition shall set forth the amount of the respective compensation, if any, for the purpose of agreement on the said amount;

b. In the same hearing or in the one following receipt of the petition, the Court shall allow the same and subpoena the concession-holder for the act of replication, by publishing the petition and the subpoena in a major circulation newspaper of the city of Caracas. Such publication shall be effected no later than 3 days following the date of the hearing on which the petition was received.

c. The replication to the expropriation petition shall deal only with the amount of compensation proposed and shall take place at the third court hearing following the date of the aforesaid publication.

d. If the concession-holder agrees to the amount of compensation proposed in the expropriation petition, the expropriation proceedings will conclude and it shall thus be declared in the sentence of the Court on the occasion referred to under subheading g. of this article.

e. If no agreement can be reached and the Court deems it advisable, it will order the appointment of experts as provided hereinafter, for the purpose of an expert accounting appraisal of the property subject to expropriation. The time shall be set in the hearing that follows that of replication for the designation of experts, one for the Attorney-General of the Republic, another for the concession-holder, and the third for the Court. In the same hearing the Court shall order that the experts thus appointed shall be notified, and such notification will be carried out within 3 days following the date of such Court hearing, and the experts shall be informed that they are to appear in Court at the first hearing following the expiration of the aforesaid term, for the purposes of accepting their posts and being sworn in. If one or more of the experts should ask to be excused or cannot be notified, once and for all, at the hearing that shall follow the hearing fixed for the acceptance of the posts and the swearing-in, the Court shall proceed to appoint the corresponding substitutes, following in such case the notifications procedure indicated above. The experts to be sworn in, whatever their number, shall produce their report in writing within 20 days of the date of the last acceptance and swearing-in.

f. The failure of the concession-holder to appear in Court for the act of replication will be equivalent to an acceptance and agreement with respect to the corresponding expropriation petition.

g. At the third court hearing following the act of replication, whenever an agreement exists or when the concession-holder should fail to appear in Court, or within 10 court hearings following the act of presentation of the experts' report, or upon the expiration of the period provided for in subheading e. hereof for the presentation of experts' report without such having been presented, as the case may be, the Court shall hand down its sentence in which the expropriation will be declared, establishing the amount of the compensation it may stipulate, and ordering the payment thereof in the form provided for in the expropriation petition.

The decision of the Court through which the expropriation trial is declared closed or the expropriation completed, shall serve the State as a title deed to the rights and assets object of the expropriation.

Article 14: In cases where the respective defendant should not agree in the act of replication to the amount of the compensation, or where the expiry of the concessions should occur pursuant to Article 1 of this law, in the expropriation petition referred to in the above article the Attorney-General of the Republic shall request the Political and Administrative Division of the Supreme Court of Justice to decree the prior occupation of the property object of the expropriation.

For the purposes of such prior occupation, the following special procedure shall be followed:

a. If no agreement is reached or if the termination of the concession should occur, the Court, in the same act of replication, shall decree the prior occupation of the assets, without the National Executive being obliged to deposit in the Court the amount of compensation offered in the expropriation petition.

b. Once the prior occupation is decreed, at the following hearing the Court shall commission a competent judge in the jurisdiction where the defendant has its main offices in this country, so that the said judge may proceed to execute the sentence and put in concession of said property the state agency designated for the purpose by the National Executive.

On the date on which such agency takes possession of said property, the hydrocarbon concessions subject to the respective legal proceedings which are not ended according to Article 1 of this law, shall cease to take effect.

The judges commissioned shall proceed to implement the measures referred to in this article in preference over any other matter. Those who should fail to comply with this obligation shall be liable to penal, civil or administrative actions as well as to any disciplinary penalties that may be considered pertinent.

Article 15: To all legal purposes and effects of this law, including the expert accounting appraisal referred to under subheading e. of Article 13 hereof, the amount of compensation for the property rights over the assets expropriated shall not exceed the net value of said assets, plants and equipment, understanding as such the purchase price thereof, less the accumulated amount of depreciation and amortization on or before the date of the expropriation petition, according to the books used by the concession-holder for income tax purposes.

From the amount of said compensation, the following deductions shall be made:

a. The value of the assets pertaining to the concessions which, in the judgement of the Ministry of Mines and Hydrocarbons, fall within the provisions of Articles 9, 13 and 15 of the Law Concerning Properties Subject to Reversion in Hydrocarbon Concessions and upon which no resolutions have as yet been dictated ordering the concession-holders to surrender them to the Nation.

b. The value of the oil extracted by the expropriated concession-holders beyond the limits of their concessions, according to the volumes established in the agreements for joint exploitation of oil resources signed with the Corporación Venezolana del Petróleo. When such agreements have not been signed, the National Executive shall determine the amounts to be deducted on this account.

c. The amount of social benefits and other rights to which Article 23 of this law refers, in the event that such benefits have not been deposited in accordance with the said Article.

d. The amounts owed by the respective concession-holders to the National Treasury and other public bodies, and any other sums which are legitimate and proper pursuant to law, except for those that correspond to income tax for 1975, which must be paid in cash.

Special Paragraph: The rights in favour of the National Treasury, as well as those of the workers against the concession-holders, which are subject to judicial debate, are excepted. Such amounts as the concession-holders may owe to the National Treasury and to the workers by virtue of the exercise of such rights, shall also be deducted from the compensation payments due, or from the Guaranty Fund referred to in Article 19 of this law.

Article 16: The payment of the compensation less deductions may be deferred for a specified length of time not exceeding 10 years, or may be paid in national debt bonus under terms convenient to the national interest, as determined by the National Executive after prior consultation with the Central Bank of Venezuela. The interest earned on such bonds shall not exceed a yearly rate of 6 per cent.

Article 17: The National Executive may in fulfilment of the compensation referred to in Article 15, deduct therefrom the amounts which the respective concession-holder may owe to the National Treasury or any other public bodies, and any other amounts according to law which, for any reason whatsoever, may not have been included in the deductions provided for in Article 15, or which may have become demandable after the date of publication of the expropriation judgement. In any case, the National Executive may impute to the Guaranty Fund referred to in Article 19 hereof whatever amounts the concession-holder may owe.

Article 18: Except as provided for in Article 23 of this law, the State shall not assume any obligation whatsoever for liabilities the concession-holders may have incurred with third parties either within the country or abroad. In cases where mortgages or preferential rights exist on the property transferred to the State under the provisions of this law, such credits shall be taken into account in determining the compensation once the deductions provided for in Articles 15 and 17 of this law have been made, subject to the same conditions under which the said compensation is to be paid to the expropriated concession-holders.

Revaluations of any kind made by the concession-holders before the promulgation of this law shall have no effect whatsoever over the determination of the net value of the expropriated property referred to in Article 15 of this law.

Article 19: For the purpose of guaranteeing the fulfilment of any and all of the obligations imposed by the present law, including those provided for in Article 17, the Guaranty Fund stipulated in the Law Concerning Property Subject to Reversion in Hydrocarbon Concessions is hereby amended as follows:

a. Within 60 days of the promulgation of this law, the holders of hydrocarbons concessions shall deposit in the Guaranty Fund, in a lump sum, the amount required in order that, when it has been added to the existing deposits in the Fund, the total shall add up to the equivalent of 10 per cent of the accumulated gross investment, as accepted for the purposes of income tax. In consequence, once the said deposit has been made, the concession-holders will be exempted from payment of the quotas provided for in the aforesaid law and in Regulation No. 2 thereof.

b. The administration of the Fund shall continue to be governed legitimately by the provisions of Article 6 of the Law Concerning Property Subject to Reversion in Hydrocarbon Concessions and of the aforesaid Regulation No. 2 thereof.

c. The Guaranty Fund shall cease to be subject to the provisions of this law once the obligations it is destined to guarantee have been fulfilled to the satisfaction of the National Executive.

d. The holders of hydrocarbons concessions may utilize national debt bonds received by them in accordance with this law, to cover in full or in part the increase of the Guaranty Fund referred to in subheading a. above.

Paragraph One: The concession-holders which accept the offer made by the National Executive within the period established therefor in Article 12 of this law, shall make the deposits referred to in this article at the time they receive such bonds.

Paragraph Two: The credit balances of the National

Treasury shall have preference over any amounts due to other public or private creditors.

Article 20: The National Executive shall carry out the inspections and examinations necessary to verify the physical existence of the assets expropriated by the Nation, as well as their state of conservation and maintenance, within a period of not more than 3 years reckoned from the date on which said assets are received.

Article 21: Acting through the agency of the Ministry of Mines and Hydrocarbons, the National Executive determine the geographical areas in which the enterprises it may establish in accordance with the provisions of Article 6 hereof shall carry out their activities; and shall ascribe to them the assets received by the State under this law and under the Law Concerning Property subject to Reversion in Hydrocarbon Concessions, including those considered as immovable assets of the private domain of the Nation. Insofar as is suitable, the aforementioned areas shall maintain the same dimensions, divisions and other specifications corresponding to the rescinded concessions.

Article 22: The National Executive and the enterprises referred to in Article 6 hereof shall have the right to continue using the assets of third parties under such terms as the National Executive may establish therefor.

The rights of way constituted in favour of the concession-holders on or before the date of expiry of the concessions, pursuant to Article 1 of this law, of the Agreement provided for in Article 12, of the publication of the judgement, or of the decision decreeing the prior occupation referred to in Article 14, shall continue in force for the benefit of the State or of the respective enterprise, for the terms and under the conditions in which the same were originally constituted.

Article 23: The social benefits of the petroleum workers stipulated in the Labour Laws and in the collective contracts are acquired rights. The amount of the social benefit accruing to each worker who is not part of a trust constituted under the Labour Laws or under the plans established by mutual agreement between the companies and their workers at the time this Law is promulgated, shall be deposited in the Central Bank of Venezuela within 15 days following the said promulgation. The social benefits shall be calculated on the basis of the worker's salary on the date on which the employer is changed, without affecting the continuity of the labour relation, that is to say, when the concessions expire or cease to have effect or when established in the written agreement. In accordance with the Labour Law the social benefits will be given to the worker when he terminates his labour relation.

The Fund thus constituted in the Central Bank of Venezuela shall be governed by the regulations enacted for such purpose, and the capital of the same may only be placed, with the authorization of the beneficiaries thereof, in safe, income-producing, high-liquidity investments. The profits produced by such investments shall be distributed in proportion to the credit balance of each worker and may be accumulated or distributed at the option of each worker. The Trust Funds constituted with the Social benefits of the worker cannot lose their essential nature and consequently do not form part of the common chattels available to creditors of the beneficiary of the Trust Fund. Workers may use the balance of their account to guarantee obligations incurred with banks or other credit institutions legally established in the country, whenever such obligations have been incurred to finance the acquisition, expansion or improvement of his dwelling, the furnishing of his home, the education of his children or the maintenance of his family's wealth.

Article 24: The workers of the petroleum industry, with the exception of the members of the Board of Directors, shall enjoy employment stability and may only be dismissed for reasons expressly provided for in the Labour Laws. In the same manner, the State shall guarantee the existing collective bargaining system and the enjoyment of the social, economic, welfare, union, and professional improvement benefits and any other benefits established in collective contracts and in the labour laws, as well as any houses, incentives, and other perquisites or allowances or fees which by customs and by the application of personnel administration rules have been traditionally enjoyed by the workers under policies followed by the companies in this matter. Likewise, the State shall guarantee the benefits of the retirement plans and their respective pensions for workers retired prior to the date of termination of the concessions as provided for in Article 1 of this law, the Agreement reached under Article 12, or the publication of judgment referred to in subheading g. of Article 13 of this law. These pension plans, as well as other plans instituted by the companies for the benefit of the workers, including the workers' Savings Funds, shall be maintained after the nationalization of the industry.

The provisions contained in the law which created the National Institute for Cooperative Education shall be applied to the Corporación Venezolana del Petróleo and to the enterprises to be established pursuant to this law.

Article 25: This law shall not affect in any way the rights transferred and the areas assigned to the Corporación Venezolana del Petróleo pursuant to provisions of Article 3 of the Hydrocarbons Law, provided it does not contradict Articles 6 and 21 of this Law. The rights which private contracting companies may hold arising from the agreements signed by them with the Corporación and published in the Official Gazette Number 1,495 (Special Issue) of 13 December 1971, shall be subject to the expropriation procedure stipulated in this Law, except that any applicable compensation shall be limited to the amount of the investments made in the block where commercial production had been determined, excluding the bonuses already paid.

Article 26: Any infractions of this law shall be penalized by the Minister of Mines and Hydrocarbons with fines ranging from one hundred thousand bolívares to one million bolívares according to the seriousness of the offence. The aforesaid penalty shall be applied without prejudice to any civil, penal, fiscal, or administrative actions which the infraction may originate, or to the administrative police measures necessary to restore the legal situation infringed or to impede infraction. The aforesaid fines may be appealed before the Political and Administrative Division of the Supreme Court of Justice, within 10 days of notification.

Article 27: The net value of the rights over the assets which shall devolve on the Nation, and the unamortized cost of the concessions, shall not be deductible for income tax purposes.

Article 28: The provisions of the Hydrocarbons Law and the Law Concerning Property Subject to Reversion in Hydrocarbon Concessions, as well as any other which conflict with this Law, are hereby revoked.

Given, signed, and sealed in the Federal Legislative Palace, in Caracas, this 21st day of August, 1975, 166th year of Independence and 117th year of Federation.

The Chairman, Gonzalo Barrios; the Vice-Chairman, Oswaldo Alvarez Paz; the Secretaries, Andres Eloy Blanco, Mazzini Mail Negrette.

Given, signed, and sealed in the Oval Room of the Federal Palace, in Caracas, this 29th day of August, 1975, 166th year of Independence and 117th year of Federation.

Let it be executed.
Carlos Andres Pérez

Countersigned, *the Minister of the Interior, Octavio Lepage; the Minister of Foreign Affairs, Ramón Escovar Salom; the Minister of Finance, Héctor Hurtado; the Minister of Defence, Romero Leal Torres; the Minister of Development, José Ignacio Casal; the Minister of Public Works, Arnoldo José Gabaldón; the Acting Minister of Education, Ruth Lerner de Almea; the Minister of Health and Social Welfare, Antonio Parra León; the Minister of Agriculture and Stockbreeding, Carmelo Contreras; the Minister of Labour, Antonio Leídens; the Minister of Com-*

munications, Leopoldo Sucre Figarella; the Minister of Justice, Armando Sánchez Bueno; the Minister of Mines and Hydrocarbons, Valentín Hernández; the Ministers of State, Gumersindo Rodríguez, Guido Grooscors, Manuel Pérez Guerrero, and Constantino Quero Morales.

The First Anniversary of the Nationalization of the Venezuelan Oil Industry

Faith in the future in 1975 paved the way for very promising results in 1977

'*I have absolute faith in the successful outcome of the state's take-over of the oil industry, after the promulgation of the law nationalizing the production and commercialization of hydrocarbons*', is what I declared emphatically in a speech from the rostrum of Venezuelan Congress on 6 August 1975.

The results of the operations of the nationalized industry in its first year have confirmed the reasoned optimism with which I assessed the future on that occasion. Concrete facts, as incontrovertibly as ever, have once again given the lie to the demagogic forecasts made by those members of Venezuelan Congress who refused to cast their votes in favour of the law which was then under discussion. Some representatives withheld their support because, in their opinion, the only valid nationalizations are ones carried out along Russian or Cuban lines. But the majority opposed it out of blind adherence to an irrational mode of thought and action stemming from the idea that it is the duty of the opposition to try to block every single initiative taken by the governing party. But, we might ask: what about the interests of Venezuela? It would appear that calculations about the future of our country are of little importance to some political groups, which are simply the victims of their own introverted fantasies.

The achievements of the nationalized oil industry have also confounded the predictions of that minority group of déclassé Venezuelans who sycophantically admire everything that comes from abroad and have thus managed to convince themselves that Venezuela and her people are physically incapable of making progress without the aid of crutches supplied by foreigners. These are the people who suffer from a pathological inferiority complex nurtured by the humiliating obeisance they insist on performing to Anglo-Saxon culture and its values. These despicable attitudes maintained by one sector of our society are further strengthened by an uncontrollable fear of each and every form of change in the status quo. These people are of course perfectly happy with their present condition; they enjoy both the material advantages of wealth and the social status ascribed to the high exclusive moneyed oligarchy.

The positive achievements of the Venezuelanized oil industry during its first testing year in existence provide a complete and welcome justification for the faith of all those Venezuelans, among the general public as well as in the government and in Congress, who were confident that the nationalization law would be a success.

The transfer of ownership and full control over the industry from foreign interests to the Venezuelan government, which was the culmination of a carefully planned, politically realistic policy, has been carried out in exemplary fashion. In the nationalized industry's specially-created mother company, *Petróleos de Venezuela* (*Petroven,* PDVSA), both management and labour have proved to be fully equal to the historic rôle they were called upon to play; they took on their new responsibilities with a highly professional level of efficiency. The technicians who took the step from private enterprise to the state company turned out to be Venezuelans who both loved their country and believed in it. Their efficiency in the service of the state company has destroyed the basis for the hasty prejudice of public opinion which had dismissed them as the incorrigible lap-dogs of the Yankees. Just like the workers in the oil companies, these technicians had signed contracts with Creole (Esso), Royal Dutch Shell or Gulf; but they had not signed away their consciences at the same time. Moreover the oil workers, despite their traditional reputation as the most combative of all Venezuela's trade unions, have also demonstrated their loyalty to their country. They have not attempted to intimidate their new employer, the state, with aggressive defiance, with disputes or go-slows to sabotage production. Now that Venezuela is in charge of her own oil, she has become an adult nation, ready to exercise full control over all her basic wealth.

A Good Balance-sheet

The achievements of the Venezuelanized oil industry during its first year in existence can best be assessed by means of the following brief summary of its activities.

1. *Production and Reserves*

Because of the government's conservationist policies, average daily production in 1976 worked out at around 2·3 million barrels, which was more or less the same as in 1975. Proven reserves remained almost intact; they amount to over 18,000 million barrels of oil and 40,000 million cubic feet of gas. Therefore, on the basis of the 1976 level of production, it can be estimated that there will be reserves of fuel in Venezuela's wells until the year 2000. It should also be pointed out that after preliminary explorations additional reserves containing large quantities have been confirmed in the Orinoco bitumen band, of which further details are given below. Significantly, in 1976 Venezuela's reserves of light crudes were tapped at a high rate with little concern for future needs. The heavy crudes, on the other hand, have always been exploited at a low rate in relation to the total reserves and their potential productivity. Thus, on the basis of the figures for the reserves still intact at the end of 1976, there will only be reserves of light crudes left for fifteen years at the present rate of production, and of the middle crude fractions for eighteen years, as against nearly forty years of reserves of the traditional heavy crudes.

The production of natural gas went up to nearly 38,000 million cubic metres, and 92% of that amount was actually used. The proven reserves of gas total more than one billion cubic metres.

The reserves in the Orinoco bitumen band are estimated at thousands of millions of barrels of heavy crudes. But the exact amount of oil in that gigantic reservoir cannot be accurately measured until fuller exploration has been carried out.

2. *Refining*

The processing of crude oil in Venezuela, to provide refined products for the rapidly-expanding national market as well as for export, plays an important part in this country's oil industry.

In 1976 the capacity of the eleven refineries in this country was nearly 1½ million barrels per day. However the refineries' production during 1976 averaged one million barrels per day, that is only 66% of their potential output.

During the year in question the Venezuelanized industry began to develop extensive projects for improvements to the present somewhat obsolete network of local refineries. The aim is to equip the refineries so that they are able to convert residual fuel and heavy crude distillates into petrol and other light crude fractions marketable at much better prices. These plans entail three major developments. First they will enable the refineries to produce the fuel necessary to meet the spiralling demand of the petrol-orientated national market. Secondly these changes will widen the range of products available for export. Thirdly they will expand the productive capacity suitable for refining the heavy crudes, the type of oil in which Venezuela is richest.

3. *Exports*

During the first twelve months after the nationalization of the industry the volume of export sales continued to grow, but not at an accelerating speed; this was due to a deliberate restriction of production in line with conservationist policies. Exports totalled 2·15 million barrels per day, which was more than 90% of the total production of oil in this country. 90% of that amount, two thirds of it consisting of crudes and one third of it of refined products, was shipped to markets in the Western hemisphere, chiefly the USA, Canada and the Caribbean.

The United States remains, as it has traditionally been, the largest market for Venezuela's oil. In 1976 that country with the dimensions of a continent soaked up 34% of our oil exports directly, plus another 15% indirectly, after being processed in the refineries of the Caribbean. Nearly 12% of our exports went to the Eastern provinces of Canada, the country which was our second largest customer for oil.

77% of the 785,000 barrels of refined products we exported per day consisted of residual fuels. Most of the exports of this type of oil went to the East coast of the United States. Venezuela supplied one third of the total volume of this fuel imported by the USA.

It should be stressed that during 1976 the government's consistently-stated policy of attempting to diversify export markets was implemented with considerable success. Many sales were negotiated directly with the governments of other countries without resorting to intermediaries. Nearly 20% of Venezuela's export sales involved non-traditional clients.

4. *Capital Investments*

The total capital expenditure of the oil industry during 1976 amounted to approximately 1,400 million bolivares (330 million US dollars), which represents an increase over previous years' investments. 87% of that sum was devoted to exploration and production, the aim being to maintain a high potential level of production and to minimize the rate at which our reserves of oil and gas are exhausted.

5. *Financial Results*

Nationalized industries in market economies have a perfectly justified bad name for running up annual deficits, which the state (or, in real terms, the taxpayer) finances so as not to lose face.

However this has not been the case with our nationalized oil industry. Venezuela has learnt the lessons provided by the experience of Mexico. *Petroven*, the state company which has administered the industry since it was wrested from foreign control, has not followed in the footsteps of its Mexican sister organization, *Pemex*. This country has steered well clear of the process of bureaucratization and politicization which has hampered the state-controlled oil industry in Mexico and in some periods has dented its prestige. To put it in concrete terms, Venezuela has not only satisfied her national pride as an autonomous state by taking full control over her basic source of wealth. *The nationalized oil industry has also earned more money for the treasury and for the nation than the industry did when it was controlled by the transnational companies which kept the lion's share for themselves.*

The tabulated financial results of the industry for the first year after its nationalization give a very good impression indeed. The total value of the industry's sales was approximately 39,000 million bolívares (9·2 billion dollars), and the nation's share, including the sums retained by *Petroven* for investment purposes, was 33,000 million bolívares (7·8 billion dollars). That works out at roughly 38 bolívares (nearly 9 dollars) for every barrel produced, which is 10% more than what the government earned in the last year in which the concession system was in force. *It is in fact the highest income per barrel obtained by the nation throughout the whole history of the industry.*

6. *Research*

Another positive achievement has been the creation of the Instituto Tecnológico Venezolano del Petróleo (INTEVEP), as a subsidiary of *Petroven* SA. This institute has already undertaken research on some crucial problems, including that of combustion *in situ* to increase the yield from heavy crudes, and the exploration and development of offshore oilfields with a view to advancing production in this area in the future. The Instituto Tecnológico Venezolano de Petróleo (INTEVEP) is destined to play a rôle of great importance in providing the technological and scientific support which the nationalized industry needs.

Venezuela, the country with the cheapest petrol in the world

Venezuela holds one world record, to use sporting jargon, which is certainly enough to make some fools and ignoramuses happy. There are figures which provide a telling comparison and show that my country is top of the league table of countries which waste their precious and irreplaceable black gold at the rate of thousands of tons per day, simply because the consumer can get it at a very low price. Here is a small table which illustrates this strange situation:

World prices for petrol and diesel fuels at the end of 1976
(in bolívares per litre)

Country	*Medium Grade*	*High Grade*	*Diesel Fuel*
Venezuela	0·15	0·35	0·10
Italy	2·43	2·53	0·76
Argentina	1·29	1·44	—
West Germany	1·59	1·70	1·49
Brazil	1·73	2·15	0·83
USSR	0·53	—	—
France	1·65	1·78	1·15

(Source: *Energy Week*: Vol. 3 of the OPEC Statistical Yearbook for 1975)

The absurdly low price at which petrol is sold inside Venezuela is not the result of a miracle of modern economics. There has never been a miracle in this sphere and I doubt if there are ever likely to be miracles in other spheres either. *Basically the state subsidizes the sale of this fuel in an irrational way, thus keeping the price lower than anywhere else in the world. This has two main results, both of which are negative for the country as a whole: (a) this encourages excessive consumption – indeed senseless squandering of an invaluable, non-renewable source of wealth, and (b) this also restricts the government's scope for putting some of the money which the owners of motor vehicles at present convert into smoke and pollution, to good use in the interests of the community which for the moment seems hell-bent on gas-guzzling.* The destructive effects of the freezing of the price of petrol are best described in direct terms, with simple figures.

In 1977 the government is subsidizing the internal market by not collecting petrol royalties worth 450 million bolívares (at 6 to 8 bolívares the barrel).

If we compare the price charged in the internal market (30·21 bolívares per barrel) with what Venezuela gets for her oil exports (65·78 bolívares per barrel), that means that in 1977, on present trends of consumption, the country will be foregoing a total of 1,818 million bolívares.

On top of that we must also add the 400 million bolívares which the companies lose each year on their running costs. That makes the total bill for the petrol subsidy in 1977:

Royalties	450 m.Bs
Difference between the value of exports and the internal market price	1,818 m.Bs
Losses for the operating companies	400 m.Bs
Total	2,668 m.Bs

If we make a projection on until 1985 at present prices, assuming that the rate of growth continues at

10–12% per year, total fuel consumption in 1985 will be two-and-a-half times that of 1977, i.e. 280,000 barrels of petrol per day, and 500,000 barrels of fuel all told.

The petrol subsidy for 1985 would therefore work out as follows:

Royalties	924 m.Bs
Difference between the value of oil exports and the internal market price	3,956 m.Bs
Losses for the operating companies	1,000 m.Bs
Total	5,880 m.Bs

For the whole of the internal market for fuels, estimated at 500,000 barrels per day, the total subsidy would amount to 8,300 million bolívares, which is the equivalent of 35% of the oils which the wells at present being tapped should be producing by then.

That does not take into account at all future rises in the world oil price, which most people agree will be between 18 and 20 bolívares per barrel in 1985, as against an average of 12·70 bolívares in 1977.

Venezuela's state oil administration is well aware that the price of oil in the internal market will have to be modified. When the government finds the right antidote to the present disease of excessive fuel consumption, it will be able to spend more on projects for the benefit of the whole community, using the revenue it will get from those who are at present wasting their money on petrol, driving about all day long in cars with massive engines, and filling up their garages with a different car for each adult member of their families.

Petrol Famine Before the End of this Century

The energy crisis predicted for the 1980s

The tenth world energy conference took place during the third week of September 1977, in Istanbul.

The setting for this meeting, attended by the cream of the world's energy experts, was the Atatürk Cultural Centre, in Turkey's largest city. The meeting brought together 3,500 specialists from many different countries, speaking many different tongues – including economists, oil engineers, electrical engineers and coal and gas experts, all of them well versed in the energy problem as a whole. The private companies in the oil business sent along some influential representatives to exchange ideas with university lecturers and with managers in the public sector, from the state-owned energy companies.

Many conflicting points of view were put forward at the meeting, reflecting differing opinions on some purely technical aspects of the energy problem. But everybody was in agreement on the key question: oil reserves are coming to an end on this planet, and mankind, therefore, is on the verge of an utterly unprecedented energy crisis.

The conclusions expressed in the final report of the 1977 Istanbul conference are the most recent of a whole series of pronouncements about the energy crisis by different groups and individuals. All these reports reach the same basic pair of conclusions, with varying shades of opinion only about the details: by an inexorable process, which will not necessarily last very long, oil reserves are drying up, and the world will be facing an energy deficit before the end of this century.

Before the Istanbul conference there had been the following influential reports: (a) the report of the OECD (the Organization for Economic Cooperation and Development); (b) a report by the 'big seven', the multinational oil cartels; (c) the WAES report (of the Workshop on Alternative Energy Strategies), famous overnight, and published in May 1977 by McGraw-Hill (New York) as *Global Energy Prospects, 1985–2000.* The whole world was shaken by the conclusions of this report, produced by a team of thirty independent experts from fifteen of the world's greatest oil-consuming countries. These are first-rate experts giving their own opinions and not those of their respective governments. The Workshop was dreamed up and organized by Carrol Wilson, a well-known professor from a university with a very high reputation, the Massachusetts Institute of Technology; (d) the precise, well-documented and fantastically clear essay by Professor Dankwart A. Rustow, entitled 'The Oil Crisis of the 1980s' (published in *Foreign Affairs,* April 1977). It is widely known that this analysis strongly influenced the team used by President Jimmy Carter to draw up the energy policy plans he put forward in the US Congress on 20 April 1977; (e) a report, mentioned approvingly on that occasion by President

Carter, produced by that highly controversial US government agency the CIA. (It is not often realized, except by students of international affairs, that the many-sided Central Intelligence Agency not only carries on a running battle against its Soviet counterpart, the KGB; it also carries out academic research of university standard in its own laboratories on social and scientific problems which are not directly linked to the fiendish business of espionage and counter-espionage.)

The facts, figures, estimates and forecasts collected together in these reports have helped to establish a pretty well-informed consensus of opinion about the energy question. The two basic conclusions of that consensus of opinion are: *(a) the world is travelling at a rapidly-increasing pace along a road which leads to the exhaustion of all its oil reserves shortly before the year 2000; (b) an acute and very wide-ranging crisis in the supply of that vital ingredient of modern society will pose a serious threat to all mankind before the end of the twentieth century.*

It is not very easy for the man in the street to assimilate the mountain of statistics on which these judgements are based. As a politician putting forward a case, and a writer seeking to inform society at large, and not as an oil expert writing for other specialists, I shall try to translate the cryptical terminology of the experts into simple language.

World energy consumption

The level of world energy consumption has risen in leaps and bounds. The increase has gone hand in hand with the growth of industrial production and the spread of a comfortable standard of living among an ever greater number of people. Figures given in the WAES report show that world energy consumption in 1900 was a mere 500 million tons a year. Fifty years later, in 1950, the total volume consumed had made a spectacular jump, to 1,700 million tons; and less than twenty-five years after that, in 1974, it had more than trebled again, to 5,800 million. Forecasts for the year 2000 suggest that by then world demand will have risen to between 15,000 million and 18,000 million tons. The reports of several different teams of great technical prestige, and the conclusions of many conferences and committees both in the government sector and in the private sector, make it absolutely clear that there is no conceivable possibility of meeting that demand. Even if the growth rate of the world economy was to remain static at 3%, the gap between supply and demand could be as much as 8%. That is an unbridgeable gap, and it could bring about a world catastrophe.

Some people have even predicted that the world energy famine at the end of the century will spark off wars of unforeseeable dimensions.

Nuclear war?

Many people who can hardly be certified as insane have suggested that the energy famine might induce the overarmed super-powers to carry out desperate attacks on oil-producing countries in the twenty-first century. The USSR might occupy Saudi Arabia or Iran or the Persian Gulf emirates and take over the lion's share of their oil wealth. Or the USA might advance into the Middle East and occupy it with paratroops; by then the US might very well have already occupied key countries inside its own geo-political sphere of influence, the Western hemisphere, for example Mexico and Venezuela. All these are possible courses of events. As even children at nursery school know, the big industrialized nations have huge arsenals of weapons which are continually becoming more sophisticated and more efficient at mass killing. From the history of mankind it is clear that when heavily-armed countries are possessed by desperate collective hysteria, or when their leaders fall prey to war fever, they use their weapons without the least consideration for humanity. However, there are two basic adverse factors which prevent such nations from breaking up the whole world order with impunity. First of all not only would the blood of those who were attacked flow freely; the aggressors too would shed a lot of blood. Few countries in any corner of the world would accept the degradations of slavery without being prepared to die in battle for their freedom, even if they were reduced to using the kind of weapons used by stone-age man. The other great obstacle to aggression is nuclear fission, which was made possible by Einstein's mathematical formulae. It is quite inconceivable that the oil-producing areas of the world could be amicably divided up between the US and the USSR on the basis of geographical proximity or convenience of exploitation. Because of their undying imperialistic rivalry, it would be impossible for either super-power to remain content with the oil reserves allotted to it by such an agreement. Neither of them would be happy with only a certain number of the militarily weak countries which happen to have oil in their subsoil, but each would be determined to get hold of all the oil reserves situated in the developing countries. Only in this way would they be able to ensure that, after exhausting the reserves of one or more countries, they could go on to exploit the reserves of those countries which still had reserves. This scenario may sound a bit like science fiction, but it would lead inexorably towards the destruction of the whole world in an atomic war. This *is not* – and, it is reasonable to suggest, *will never in the twenty-first century be* – contemplated by the rulers of the USA or the USSR. In the future those who are in power in Washington and Moscow will be as conscious as those who are now in power of the fact that, if the deadly energy contained in the millions of megatons stockpiled in their atomic arsenals was let loose, it would leave nothing alive – not even the goat which refused to die on Bikini.

Alternative sources of energy in place of oil

Gas: This is not a viable substitute for oil. The proven world reserves of this fossil fuel are similar to those of oil, and will run out, according to reasonably precise forecasts, at about the same date.

In 1977 liquid gas accounts for 20% of world energy consumption. It is not feasible to increase this already significant proportion in the near future. The main difficulties preventing the wider use of gas are the problem of transporting it and the shortage of processing plants. There are long distances between the main production areas in the USSR and the OPEC countries of the Middle East and North Africa and the key consumer markets in the Western world. In addition the present network of gas processing plants is not big enough to cope with greatly increased quantities of natural gas. For these reasons, as the WAES report concludes very forcefully, by the year 2000 the industrialized Western world's requirements in terms of imported gas will be far greater than the facilities for transporting and processing it. Reasonable estimates from reliable sources state that, in order to overcome these two problems – bottle-necks in transport and in processing – a tremendous effort and huge capital investment in tank-ships and plant will be required.

The cost of a proper attempt to exploit the gas reserves of OPEC has been worked out by M. Ait Lanhoussine, the vice-president of Sonatrach, the Algerian state enterprise, with a precise estimate, which at the time of writing remains unchallenged. One thousand great gas tankers and one hundred processing plants (each of which would cost 1·5 billion bolívares, at, say, 4 bolívares to the dollar) would be needed for this purpose.

Coal: This is not a substitute for oil; it predated oil and was superseded by it about sixty years ago, when the industrialized world discovered, during the course of World War I, that another natural source of energy, namely petroleum (like coal, a fossil fuel), provided a more efficient method of running engines for factories and for mechanized weapons on land and sea and in the air. In addition oil was very, very cheap. The imperialistic super-powers of that time, those great exploiters, the United States and Great Britain, paid for their oil a miserably miserly price to the hungry colonial countries of the Middle East and the semi-colonies of Latin America – Mexico until 1938, and

Venezuela until well after that. Coal-burning machines were converted so that they could use what was for decades a very cheap liquid fuel, black oil. The result was that the landscape in large areas of the industrial zones of Great Britain and the United States was filled with mammoth relics of scrap iron – the iron and steel remains of machines which were abandoned because they could not be converted to burn anything other than coal. In the golden age when oil held sway, coal suffered the cruel fate of an overthrown ruler and was forgotten. For decades it was kept buried deep in the mines, and was only brought up to a very limited extent, for example in poorer countries like Spain and Belgium. By a strange coincidence in both these countries the coal was actually dug by Spanish workers, who, both in Asturias and in the Liège area, gave their utmost for coal, exhausting their bodies and racking their lungs with silicosis.

Because of this half-century of limited mining the world's coal reserves are at present considerable. The report of the tenth world energy conference, held in Istanbul in 1977, provides the best and most complete account of these coal reserves, their energy potential, and the ease, or otherwise, with which they can be commercialized, etc.

The estimated total of these vast reserves is 10,000 billion tons. However only a fraction of that amount is considered to be exploitable, given the present price of energy and present techniques – a mere 600 billion tons or so, which is the equivalent of 400 billion tons of oil.

The world's coal mines are dotted about the globe just as arbitrarily as the great oil and gas-fields. Three large countries, the United States, the USSR and China, possess more than two-thirds of total reserves.

There are several serious problems which hamper the collective use of the world's coal reserves. The two main problems are the difficulty of transporting it and the resistance of consumers against using it.

The transport problem is particularly remarkable in the United States. That country's reserves are so big that it could become the world's greatest coal exporters, but this is prevented by the decay and near-collapse of its railway network, which has literally fallen apart because of disuse. The mines themselves have not been touched for several decades. The WAES report even details the enormous sums which would have to be invested in the USA to get the mines going again and rebuild the railway system in order to quadruple the present figures for coal production by the year 2000. The modernization and reactivation of mines will require 32 billion dollars, and the transport system, that is the railways, a further 86 billion dollars.

These astronomical sums cannot be invested by the government alone. The cooperation of private investment capital will be needed on top of government expenditure. This form of joint action by government and private capital has been advocated very vigorously by President Carter, who is also a strong supporter of the idea that energy derived from coal should be used to cover part of the oil deficit. It has however been pointed out that the exhortations of the US president, even if they are backed by legislation, will not persuade that country's capitalists to invest millions of dollars in the coal industry until that product's price is comparable, on a competitive basis, with other sources of energy. The behaviour of the investors, concentrating on their own profits at the expense of the national interest, should not really surprise us if we remember that the US is the archetypal market economy. But the cautious response of Wall Street investors to offers of holdings in coal companies will not last much longer, because it is very easy to predict that, within a few years, the price of a barrel of oil will no longer be 12 dollars as in 1977, but will be somewhere between 20 and 30 dollars.

The second obstacle which coal will have to overcome before it can fill part of the gap between the supply and demand for oil stems from the very bad image it has among consumers. It is a dirty fuel which pollutes everything, spoils any landscape it occupies, makes those who use it feel sick and provokes constant hatred from all ecologists. It is not easy to forget the sustained campaign in the media run by the multinational oil companies for decades to prevent coal from competing with oil. In every corner of the earth people think of coal as one of mankind's worst enemies. The experts who wrote the WAES report go so far as to hypothesize that people's hostility to coal might lead by the year 2000 to a paradoxical situation in which there would be surplus production of coal which the world was not prepared to use even when its appetite for energy was unsatisfied. It is technically possible that world coal production at the end of the century will be three times what it is now. By then the oil deficit would be one billion barrels, but there might still be a surplus of 350 million tons of coal which no one would buy.

Coda: Final Reflections on a Personal Note

'He de amarte tan fuerte que no va más,
y el amor que te tenga, Venezuela,
me disuelva en ti.'
(Antonio Arráiz Quiero estorme enti.)

I am writing the final pages of this work on oil in Venezuela and the world in September 1977. I do so with a very special feeling of anticipation.

February 1978 will be a month of momentous anniversaries for me; I will complete fifty years of active political life and I shall also celebrate my seventieth birthday, since I was born on 22 February 1908. During this long career stretching over half a century I have devoted considerable attention to oil problems almost every day.

I governed Venezuela for eight years and the administrations over which I presided put into practice my ideas for the Venezuelanization of this country's oil, and also my dream that the small oil-producing countries of the world should no longer be outcasts deprived of the benefits of their own natural resources.

During my periods in office, with the government departments in the hands of men as determined as I was to persevere with revolutionary nationalist policies, Venezuela led the way towards radical changes in the relationship between the oil companies and the state, and in the organization of the industry across the whole world.

In 1946, when I was this country's head of state for the first time, Venezuela, following the dictates of sovereignty, unilaterally established the 50–50 relationship, the equal division of the profits of oil extraction between the concession-holding (oil-producing) companies and the state. This formula became known as 'the fifty-fifty' in international oil slang and spread like wildfire to the Middle East and all the other oil-rich areas of the world. For many countries which received only a pitiful proportion of the prodigious wealth extracted from their subsoil, this meant the beginning of a process by which the value of their product was increased and control over its exploitation was wrested away from the transnational companies and their sleeping partners, the governments of the industrial nations of the capitalist West. In 1960, when I was president of this country for the second time, Venezuela became one of the leading pioneers and standard-bearers of OPEC (the Organization of Petroleum Exporting Countries). OPEC represents the bravest and most successful attempt yet made by countries that are rich in raw materials but weak in military terms to defend themselves by means of a counter-attack to throw off the shackles imposed by the military and industrial strength of the great Western powers. For the first time in the modern era a limited group of under-developed or developing nations was able to coordinate its policies and thus overcome its inherent weaknesses to form a solid united front to impose its terms for marketing its product on the multimillionaires of the financial centres of New York, London, Paris and Berlin. In one crucial sector of the highly mechanized industrial economy of the mid-twentieth century, the implacably omnipotent purchasers of this raw material were no longer able to fix their price; from now on it was to be fixed by the producers of crude oil. From the dirt-cheap level at which it had stayed for decades, the price of oil at the wellhead then jumped in ten years to twelve US dollars a barrel.

I have not been a mere spectator during the course of these spectacular changes in the politics of oil which have taken place in Venezuela and the rest of the world in recent decades, the importance of which there is no need to restate. I have in fact held key positions in the country which set these changes in motion. It has been an all-embracing pursuit for me, combining vision and action, because oil is one of the most important factors in modern world history, and the very hub of political and economic life in this my native country, to which I have given my utmost.

These remarks should help to explain the way I write about oil both in this book and in its predecessor, *Venezuela, Política y Petróleo,* written over twenty years ago (first edition 1955; six subsequent editions).

In speaking and writing about oil, and in taking government decisions on this subject I have tried not to act as an expert. My idiosyncratic approach to the subject has led me not to take up the objective impartial standpoint of an economist, a historian or a political scientist. I have not used impersonal academic tools to examine the oil question but rather my own passionately committed personal vision. The destiny of my homeland – a small country with an important place on the unpredictable map of world hydrocarbon deposits – has always been very closely linked to oil. It should be clear therefore that everything I have said about oil in countless speeches, lectures and official statements, and everything I have committed to print on the subject – indeed every contribution I have made to the struggle to give Venezuela full control over her own oil – has not been just for the record; it has all been an expression of my actual day-to-day preoccupations. I have suffered frustration and despair and have also enjoyed personal triumphs. I have had to face periods of deep discouragement of the kind which makes even the most determined men pass many sleepless nights, and also moments of complete euphoria complicated by the pain that one feels when one's happiness is too great for one's capacity to register emotions – that happened on the day on which Venezuela took complete control over her oil industry. The whole long process by which Venezuela has achieved independence in economic terms has been bound up in every conceivable way with my extended career between 1928 and 1977.

As I come to write the last paragraphs of the varied history of the oil industry in Venezuela, interwoven with my own nationalistic passion for my country, I can see no clouds on the horizon for this country, which has been the focus for my first and most consuming emotions. Venezuela has now achieved freedom on a permanent basis, with a representative political system and democratic form of society and institutions, which, though still imperfect, can and will, I can state with complete confidence, reach an ideal condition in the near future. The absolute conviction with which I make these statements has a firm logical basis. We will achieve an improved form of democracy because that is the unshakeable aim of all Venezuelans with a sense of moral purpose, and of the country's majority party, the most powerful political organization in Venezuela, Acción Democrática, the party which I myself founded four decades ago in 1937. Venezuela is now no longer to be treated as a semi-colony in economic terms under the thumb of powerful foreign capitalist groups. Venezuela used to be little more than a reserve of raw materials for the industrialized nations; population had a very low cultural level, because its rulers had no idea of national responsibility and hindered the growth of an educational system, partly for lack of sufficient revenue because the lion's share of her natural resources was greedily snapped up by the coupon-cutters of Britain and the USA. Venezuela is a nation with a democratic government and democratic political customs: she is one of the emerging nations still undergoing development, and she supports the non-aligned philosophy of the Third World and its just demands. But, because of her level of economic development and her national unity in terms of language, culture and religious traditions and beliefs, Venezuela should be distinguished quite firmly from countries elsewhere in the world which still have to contend with situations inherited from colonialism in recent periods, which has left behind tribal rivalries, and divisions because of heterogeneous religious traditions or a great diversity of local languages or dialects, to impede progress. Venezuela will go on getting a good price for her oil, which will last an indefinite length of time. There will therefore continue to be an influx of foreign exchange earnings to help out the national economy and the finances of the government. That will ensure a steady advance towards a situation in which the whole population of Venezuela will enjoy economic security and well-being, and there will be a high level of culture for all.

The economic situation of Venezuela is prosperous, and her prosperity is a source of amazement for the rest of America and for the whole world. Whatever their origins, people express surprise and often ill-concealed envy when they learn about the basic statistics of our prosperity. Here is a short selection in abbreviated form, and in round figures. In 1976 the Gross National Product of Venezuela was 68,000 million bolívares (at 1968 prices): that works out at 15,000 million dollars. This figure for the GNP makes Venezuela the country with the highest income per capita in the whole of Latin America, at 2,650 dollars. The per capita income of the countries traditionally known as the 'big three' of Latin America is lower than that of Venezuela. In Argentina the average per capita income was only 1,500 dollars, in Brazil 1,170 dollars and in Mexico 1,200. The rate of growth of Venezuela's GNP continues to increase year by year, and in 1976 it rose by over 13% in comparison with the previous year. In the jargon of economists, who use an arid vocabulary full of clichés, the Venezuelan economy is 'behaving well'. (While I am on the subject of the pompous and occasionally anthropomorphic jargon used by economists, let me make an exception of one of their number, John Kenneth Galbraith, who has perhaps the greatest number of followers across the world as a whole. Two of his books are certainly not suitable reading matter for the serious-minded fools whose jaws are locked fast by bitter personal crusades against all kinds of humour. The first is *Economics and Laughter* (1971); the other is his most recent work, *The Age of Uncertainty* (1977)).

In her present situation in the world, Venezuela, as the leading oil exporter in the Western hemisphere, will always be a moderating element, a standard-bearer for international justice and an outright enemy of all warlike adventures. Venezuela's strong commitment to these positions is always backed up by her history and by the underlying characteristics of her people. Venezuela totally rejects the idea that OPEC is entitled to use its very powerful economic weapon to exert political pressure and even in some cases to blackmail third parties. We believe in dialogue and discussion, in civilized forms of debate, not hasty retaliation. As a country which has suffered flagrant injustices and poverty at the hands of the high and mighty in this world – transnational companies and imperialistic governments – Venezuela will be able to distinguish between markets with strong purchasing power and those in weaker economic positions. That is why Venezuela will not change her well-established position as the main promoter inside OPEC of the idea that a substantial part of the high revenues now enjoyed by the member countries of the Organization should be set aside to finance development and, at the very least, to mitigate the social tensions of the Third World countries without oil, including those of Latin America. Because there is no religious fanaticism in Venezuela and the majority of her inhabitants are Catholics who tolerate the divergent beliefs of members of other religions, Venezuela will not support those who might be tempted to use their oil-producing potential to try to impose their own patterns of ideas and religious worship on other countries. To maintain this immutable position Venezuela can not only point to her position as one of the main suppliers of energy in liquid form to the USA, the most powerful country in the world, Venezuela can also make use of her prestige as the trail-blazing pioneer in the process through which the producing countries were set free from the tight leash on which they were held for so long. But the best credentials which Venezuela can use for upholding this just and far-sighted position are her own national characteristics as an open, democratic egalitarian society without any kind of privileged class and without a trace of political oppression or racial or religious discrimination. The statements of a nation of this kind enjoy a warm reception across the whole world, and I do not say so out of a brash form of patriotism, but because this is the truth.

Those Venezuelans who are not blinded by emotional prejudices recognize, with deep-seated national pride, that the rulers of this country have managed to channel the proceeds of the increased price of oil towards useful purposes. In 1977 the economy grew at a considerable pace; a particularly promising sign has been the significant recovery made by agriculture after a long period of stagnation which had held back our development. The country infrastructure has been expanded and consolidated, so that it has now become fully equipped. The road system is growing in a carefully planned and organized way, and so is the network of water mains and buildings to provide facilities for education and public health. Industrial growth can be said to have taken off fully, and to have got beyond the unavoidable first stage in which national industry is confined to import substitution; we are now beginning a new stage in which a much more varied pattern of growth is possible, with the export of Venezuelan manufactures to markets in neighbouring regions or overseas. Unemployment is running at such a low level that it is no exaggeration to say that our country is close to that longed-for goal of modern society – full employment. The purchasing power of the consuming public is high and the shopping centres are always crowded in spite of inflation – and the rate of inflation in Venezuela compares favourably with that in other countries in Latin America and elsewhere in the world at a similar stage.

However I would be taking up a blind, unrealistic and totally irresponsible position if I were to suggest that Venezuela today is a kind of idyllic tropical version of Alice's Wonderland, or that it was likely to become so in the near future.

All Venezuelans with a sense of responsibility have a very different attitude from that one. Most of our ideas can be summed up in the following diagnosis of this country's real problems, which require the energetic measures I outline in order to stave off disaster.

I myself lay claim to a frontline position in this battle for our country's future.

The two basic weaknesses which must be cured are the government's absurd tax system and the unjust distribution of income – of the material wealth created in this country.

Oil accounts for roughly 70% of government revenue. Seventy céntimos out of every bolívar collected by the government come from the oil industry. By nationalizing this industry Venezuela has broken free from the many tentacles of the capitalistic oil cartels, which are known throughout the world as 'the big seven' or 'the seven sisters'. But the government, and hence the economy as a whole, still gets its life blood from a finite, non-renewable natural product. Even though the production and marketing of oil has been nationalized, both government and nation continue to depend on oil to a pretty alarming extent.

It has been argued above, with ample statistical justification, that the rate at which wealth is generated is high in Venezuela, as high as anywhere in Latin America. The way in which this fantastic volume of money is divided up between the people of this country is, quite simply, revolting because it is so inequitable. The lion's share of these riches is enjoyed by a tiny minority of the population. The World Bank's latest figures show that the distribution of income in Venezuela at present runs dead against any idea of social justice:

Percentage of national income going to the bottom 40% of the population of Venezuela	7·9%

Percentage of national income going to the middle 40%	27·1%
Percentage of national income going to the top 20%	65·0%
	100·0%

The heavy dependence of the government on oil revenue and the distribution of income in such a shockingly unjust way must be modified by a comprehensive tax reform. The exorbitantly fat profits obtained by the oligopolistic economic groups in a variety of different activities must be taxed at a proper rate. In this way the government will be able to begin to break out of the stranglehold imposed by its dependence on oil as its sole important source of revenue. It will also be in a better position to redistribute some of the money which previously went to a tiny group of people in the interests of the majority which needs a better standard of living. There can be no reasonable objections to this change towards a rational and just tax system. That is in fact what all modern countries in the West have done.

I have described the positive side of Venezuelan society in 1977, complete with a few disturbing signs. In the following paragraphs I will try to summarize some aspects of the other side of the picture, which I find both negative and deeply worrying; I shall not mince my words. I find that I cannot reconcile cowardly, evasive half-truths with my conception of patriotism, which, like that of José Martí, combines duty and personal suffering.

Venezuela clearly faces the possibility of going

rotten, or even disintegrating. A country with only twelve million inhabitants cannot support a government which spends 55,000 million bolívares (or 15,000 million US dollars) a year with impunity. Venezuela is a country full of extravagant spendthrift *nouveaux riches*; there is a crazy collective hysteria which drives people to spend their own money and that of others; Venezuela is thus a kind of working model of the ultimate consumer society, not a part out of place, not a screw loose. The pretentious facade, which announces that this is the Latin American country with the highest per capita income, and one of the ten countries in the whole world with the largest holdings of strong currencies to support its own monetary system, conceals the dramatic fact that this is a poor rich nation. Our system of values has suffered a grave distortion. The possession of money and the custom of throwing it about with a showy vulgarity, are the passport to prestige, ostentatious symbols of high status. Whole networks of contractors (I was almost tempted to say mafias) devote themselves to robbing the taxpayer with cynically good-natured abandon, by trying to make shady money out of the government or to buy the services of corrupt government officials. Both from outside the government and by making use of their official stooges who are sucking the nation's life-blood, these racketeers block the government's attempts to defend and enhance the quality of our country's most permanent resources – its people. They use the powerful machinery of mass communication – not only the press but also, yet more insidiously, the radio and television – to preach a whole cult of gigantic expansion. By a devastating piece of brainwashing they convince the country that its public investments should all be on a massive scale involving many millions of dollars at a time. On these investments there is a lot of spare fat, which is not the case with investments designed to expand the capacity of our schools, to improve the public health service, to provide better credit facilities for poor industrialists and farmers, or a more effective agrarian reform, better and more efficient public services, or cheap housing for the lower paid.

This conspiracy of multimillion dollar rackets against the national interest is beginning to contaminate all levels of Venezuelan society. Venezuela has now developed its own version of the Mexican 'mordida', a cut taken by government officials, who, though previously incorruptible in Venezuela, have now started to demand regular bribes and to sell small-scale favours.

There is a close and revealing relationship between rises in oil prices and the irregular growth of public expenditure, which has left the way open for government revenues to be wantonly squandered. Around 1970 these began to get much worse, although it is only since then that they have begun to take on really alarming proportions. In less than five years the national government's budget went up from 14,000 million bolívares to 45,000 million. That enormous jump might have turned out to be a fatal one for the country.

Now that we have confirmed this diagnosis of the moral and ethical crisis facing Venezuela and located the main sources of the infectious bacteria which are spreading through the government and the country as a whole, there is still time to start on a cure for our present trouble. Enlightened Venezuelans have the determination to apply a medical cure; that is why we can confidently express a rational brand of optimism.

It is now quite obvious to everyone who is aware of the nation's interests that there must be no delay in combating the shameful vice of administrative corruption until it is completely stamped out. The government which is in power at the time of writing has responded to public outcry on this subject by putting forward in Congress a bill which will give the government effective methods for dealing with those involved in embezzling government funds. It provides for quick, harsh penalties for those who make a living out of deceiving the public and those who take a cut out of government revenues. That bill to protect government funds and strengthen public morality will become law in 1978. Acción Democrática will get it through Congress, either with the support of other party groupings or by using its own majority. The Contraloría General de la Nación is already carrying out its normal task of exercising preventative supervision over government expenditure day by day; it draws attention to faulty administrative practices and totally illegal ones. The bill at present under consideration will widen its sphere of action and give it greater authority.

Moreover, the idea that public expenditure should be rationalized, measured and controlled has gained ground visibly in Venezuela. Ambitious plans to cut corners in the development process – plans which are both patriotic and ambitious – are in no sense incompatible with the two classic axioms of good government, which will remain true unless some really earth-shaking new ideas are developed. The two basic prerequisites for any government project are, first of all, a secure financial basis, and secondly a reserve of human resources large enough to set each project in motion and make it work.

It is highly significant for Venezuelan society that this judicious policy is inspired by the awareness of those in the government and their supporters among the general public, including some who are genuine democrats but do not belong to any party. These ideas have not been imposed by any of the political groups from outside the system, namely those which form the wide spectrum of tiny groups which shelter under the umbrella of Marxism (some of them are Leninists and others Maoists, but all of them are firmly attached to Castro's Cuba, which has turned out to be a breeding-ground for economic disasters and a replica of Stalin's bureaucratic totalitarianism).

In the last general election in this country, in 1973, there was only a tiny flurry of votes for each of these groups in and around the red sector of the political spectrum, mere islands in the crumbling archipelago formed by what I like to call the knights errant of the left. The biggest of the various competing anti-establishment parties in the 1973 elections was the MEP (Movimiento Electoral del Pueblo) which calls itself the Venezuelan Socialist Party. This is a breakaway group from Acción Democrática, started by a divisive group which has failed in political terms and been left behind by the march of history. In 1969 the MEP made a significant showing, but the 1973 elections revealed that it is an artificial grouping which has lost its footing and is liable to collapse. Its presidential candidate obtained a mere 210,513 votes, 5·2% of the total votes cast. The most successful of the parties with an avowedly Marxist stamp in 1973 was the MAS (Movimiento al Socialismo), a breakaway group from the Communist party, which uses the vocabulary of democracy but is still constrained by hard-bitten dogma; it still believes in the use of armed violence as a legitimate way to reach power, even though elections in this country are both clean and free. This party turned out 232,756 supporters in the 1973 elections, 5·9% of the total number of votes cast. The Venezuelan Communist party (PCV), which this very month (September 1977) will celebrate its fortieth anniversary, got 28,403 votes in the presidential elections and 49,455 votes in the Congressional elections. The proportion of the total vote which these figures represent is so small that it is of interest only for statistical purposes – respectively 0·68% and 1·19%. (Thirty years ago the PCV had an equally insignificant following, which shows that both then and now the Venezuelan people is fervently nationalistic and fundamentally libertarian; it rejects a party which is controlled from Moscow, with political goals akin to those of totalitarian Stalinism, which involve strangling public liberties and trampling on human dignity – their model for economic development, the Soviet Union, after sixty years is every bit as unattractive as the framework proposed by classical capitalist economies). In the 1947 general election, the first really authentic free elections ever held in Venezuela, the presidential candidature of the most famous and longest-surviving leader of the PCV got only 36,514 votes, attracting a dismal 3% of the electorate. The MIR (Movimiento de Izquierda Revolucionaria), the loudest and most offensive exponent of the Marxist-Leninist-Castroist faith, polled in the 1973 elections the insignificant total of 22,552 votes in the presidential elections, a mere 0·54%, and 42,106 votes, or 1·01%, in the Congressional elections. Two other tiny groups with ill-defined political programmes and erratic records are the URD (Unión Republicana Democrática) and the FDP (Frente Democrático Popular), both of which have now definitely missed the fast-moving bus of history. The support won by these two groups in those 1973 elections shows that this remark is not a hasty subjective comment. In 1973 the URD polled 126,000 votes in the presidential elections and 132,780 votes in the Congressional elections (that is 3·06% and 3·20% respectively). The FDP polled 32,883 votes in the first elections and 51,347 in the latter, a mere crumb of 0·79% and 1·24% respectively. These two non-Marxist mini-parties put together managed less than 4% of the total vote and another 11% went to the heterogeneous ragbag of different varieties of our own home-grown and home-spun versions of left-wing Socialism. The other 85% of the votes – more than three-quarters of the total electorate – was shared between Acción Democrática and the Social Christian party COPEI; these two parties are known in left-wing rhetoric as 'the establishment parties'. 85% of the votes freely cast on that occasion went then to the two mainstream elements in the Venezuelan party system: Acción Democrática got 46·54% of the total number of votes, with 2,128,161 votes in the presidential elections. It is reasonable to suppose that the results of these recent elections will be repeated with not too great variations in the forthcoming presidential elections of 1978, and that Acción Democrática is still the majority party, with a very strong hold over the Venezuelan electorate. That this is generally accepted is demonstrated by the fact that Acción Democrática now has a new nickname, which has a universal meaning, not merely a political one; it is known as 'the people's party'.

Despite the powerful position enjoyed by their party, the leaders of Acción Democrática have not fallen victims to overweening pride. The AD leaders still keep the promises made when the party burst on to the national political scene in such a prophetic manner forty years ago. They refuse to turn AD into the vehicle of an exclusive, monopolistic single-party system. If you scratch the surface of people who proclaim themselves to be national saviours who can exercise complete control over the country's destiny, you will soon find naked totalitarianism underneath. Democracy should involve the enthusiastic collaboration of everyone in the community who is fit and able to contribute, not just a small group to which either some supernatural power or alternatively the forces of history (that depends whether the group happens to be motivated by religious sentiment or by the philosophy of determinism) have assigned a sacred mission. In its attempts to create a better Venezuela, Acción Democrática will seek the active support of the best of the nation's inhabitants on every possible occasion.

I still maintain the instinctive confidence in the

Venezuelan people which has inspired me throughout the whole of my long and varied half-century of public life. I am convinced that the party I founded in 1937 will continue to channel and unite this country's best impulses, and will contribute towards the formation of a more enlightened nation with a greater sum total of happiness for all its inhabitants.

That was the vision of destiny I glimpsed at dawn on 20 September 1977, with the crash of the curling waves of the Caribbean on our coastline ringing in my ears, and the first rays of the morning sun just touching the horizon. This is my Venezuela, its shores bathed by the Caribbean and lit by the warm sun – my beloved native land, which is, as it always has been, the light of my life, and will remain enthroned in my heart until I breathe my dying breath.

Rómulo Betancourt

Naiguatá, on the coast near La Guaira.
20 September 1977.

Appendix I
Rómulo Betancourt

by Miriam Hood

Rómulo Betancourt was born on 22 February 1908 in Guatire in the state of Miranda in Venezuela. He came from a poor background and went to work at the age of fourteen in order to pay for his own schooling. He qualified for university at the Liceo Caracas, where the headmaster at that time was Rómulo Gallegos. He entered the Law Faculty in 1928, and soon took an important part in plotting and leading the popular rebellion against the dictatorship of Juan Vicente Gómez headed by students. Betancourt was among those who shouldered arms and attacked the San Carlos barracks in Caracas, and, as a result, was sent to prison and afterwards to a long exile. He came back to Venezuela in 1936 after Gómez had died. When anti-democratic repression began again in 1937, Betancourt went underground for three years, and led a very active, but dangerous existence. At that time he formed a dynamic group of workers, peasants, schoolteachers, students and professional men, which made a two-pronged attack on the existing political spectrum against the reactionary right and against Soviet-inspired Communism. Out of this underground movement's propaganda and organization arose Venezuela's first modern political party, Acción Democrática. Betancourt became president of a provisional government in 1945 after the overthrow of the previous régime which denied the people its right to vote, and administered government funds corruptly. In Betancourt's three years of power were laid the economic, political and social foundations of present-day Venezuela. He then presided over the first direct presidential election (in Venezuelan history), with universal suffrage and secret voting. Rómulo Gallegos won the election, but was removed from power by a military coup on 24 November 1948. It was Betancourt who took the lead again in the struggle carried on by his party and by other political groups against the dictatorship which was set up after the coup. When this régime had been laid low, in 1958, free elections were held, and Betancourt became president again for the period from 1959 to 1964. He got 50 per cent of the votes in a three-cornered fight against Rear Admiral Wolfgang Larrazabal, backed by the Union Republicana Democrática and the Communist Party, and Dr Rafael Caldera, leader of the Social Christian party, Copei. Betancourt's five years in power were fertile in political, social and administrative achievements, despite constant pressure from reactionary interests and from Communists, who combined to form the opposition to his government. In March 1964 Betancourt handed over power to Dr Raul Leoni, who had been elected in a six-sided contest, and immediately left the country, thus making a dramatic break with the bad old Venezuelan tradition by which presidents, after leaving office, interfere with policies of their successors. In the United States Betancourt has been made an honorary doctor of law by Harvard and Rutgers and by the University of California, and has lectured at several other universities. He has travelled in South-East Asia and all over Western Europe; the only capital city in the Communist World which he has visited is Budapest. In May 1976 he was chairman of the meeting which took place in Caracas of the leaders of the Socialist and Social Democratic parties of Western Europe and their counterparts in Latin America and the English- and Dutch-speaking countries of the Caribbean. As a former President of the Republic, under the Constitution Betancourt holds an honorary life seat in the National Senate.

Victor L. Urquidi, who contributes the prologue to this book, is one of Latin America's most distinguished economists, and a leading member of the Club of Rome. He is President of the Colegio de México, the most famous centre of academic research in his country.

Appendix II
Books and Articles of Rómulo Betancourt

by Miriam Hood

En las Huellas de la Pezuna, Santo Domingo, 1929.
Con quien estamos y contra quienes estamos, San José, Costa Rica, 1931.
Una República en Venta, Caracas, 1937.
Problemas Venezolanos, Editorial Futuro, Santiago de Chile, 1940.
Trayectoria Democrática de una Revolución, Imprenta Nacional, Caracas, 1948 (Presidential papers for Betancourt's 3-years in power, 1945–8).
Panamericanismo y Dictodura, Mexico, 1953.
Pensamiento y Acción, Mexico, 1954.
Venezuela, Política y Petróleo, Fondo de Cultura Económica, Mexico, 1956 (second and third editions, in 1967 and 1969, by Editorial Senderos, Caracas).
Hacia América Latina Democrática e Integrada, first edition (June 1967) and second edition (August 1967) by Editorial Senderos, Caracas; third edition by Taurus Ediciones, S.A., Madrid, 1969.
La Revolución Democrática en Venezuela, (A collection of presidential papers from Betancourt's period in office between 1959 and 1964).
Venezuela Duena de su Petróleo, Ediciones Centuaro, Caracas, 1975 (14 editions).

Ensayo sobre América Latina y sus relaciones con el resto del Mundo (a speech given at the opening session of the meeting of leaders of the Socialist and Social Democratic parties of Western Europe, and their counterparts in Latin America and the English- and Dutch-speaking countries of the Caribbean, for which Betancourt was in the chair, 22 May 1976) in *Resumen*, Caracas, 30 May 1976.